Euclid's *Elements*
Books I - VI
With Exercises

Instructor's Copy

Kathryn L. Goulding

The text of Euclid's *Elements* is from Dr. David Joyce of Clark University and is used with his permission.

Copyright© 2004 Kathryn L. Goulding
All Rights Reserved
ISBN: 978-0-692-92595-9

Preface

"There never has been, and till we see it we never shall believe that there can be, a system of geometry worthy of the name, which has any material departures [...] from the plan laid down by Euclid." – Augustus DeMorgan, Oct. 1848, quoted by Sir Thomas Heath in *The Thirteen Books of Euclid's Elements* Vol. 1, Introduction and Books I, II. Preface, p. v.

"Euclid once superseded, every teacher would esteem his own work the best, and every school would have its own class book. All that rigour and exactitude which have so long excited the admiration of men of science would be at an end. These very words would lose all definite meaning. Every school would have a different standard; matter of assumption in one being matter of demonstration in another; until, at length, Geometry, in the ancient sense of the word, would be altogether frittered away or be only considered as a particular application of Arithmetic and Algebra." – Dionysius Lardner, in the year 1846, from the preface to his edition of the first six books of Euclid's *Elements*.

The most cursory glance at a selection of modern geometry texts confirms Lardner's prediction of over 150 years ago. There is no common standard for axioms; what one text assumes another proves. Definitions, some of them pitifully inadequate, are scattered throughout a text. The study of geometry, for the beauty of its "rigour and exactitude" has disappeared from the curriculum of high school students. Instead, students study geometric facts, practice algebra applied to lengths of lines and degrees, and prepare for standardized tests.

The aim of this work is twofold. First, it was written for the student who wishes to approach the study of geometry from an axiomatic standpoint, who wishes not only to learn geometry but desires to develop his reasoning skills as well.

Second, it aims to give the interested teacher, even one unfamiliar with Euclid, enough tools that he or she feels comfortable teaching geometry from the *Elements*, a treatise most college graduates have never seen. But in spite of any challenges rediscovery of the *Elements* may present, the rewards for both student and teacher are great. One of the delights of Euclid's *Elements* is that, while it uses the same deductive reasoning used in higher mathematics, its subject matter is easily accessible to a high school student; the mathematics teacher exposed to it for the first time will find much pleasure in its elegance.

One small caution is in order here. Mathematically, the *Elements* is not without problems. In the section of Teacher's Notes, I have mentioned some of these difficulties. For the most part, however, and certainly from a student's point of view, the first six books of the *Elements* are a complete and logically coherent presentation of plane geometry, the best, in fact, that has ever been written.

The teacher or student desiring a detailed exposition of the *Elements*, including discussions of alternate definitions and proofs, comments on the original Greek, and much data of historical interest, is encouraged to consult Sir Thomas L. Heath's translation and very extensive commentary on the *Elements*. It can be purchased from Dover Publications at modest cost. Also, at this writing, there is an excellent web site by Dr. David Joyce of Clark University containing all thirteen books of the Elements and some very interesting guides to each proposition.

Most students today go through high school (and college, too, for that matter) without a single course in formal or informal logic, deductive reasoning, or argumentation. Geometry is the easiest way to introduce the student to good formal reasoning while having the benefit of simultaneously teaching him a subject, which, if he wants to call himself educated, he needs to know. Taught well, it can lead students to an appreciation of the deductive reasoning processes that are at the heart of all mathematics. It is the rare student who cannot profit from its study.

Kathy Goulding
Harvard, Massachusetts
2004

To the Teacher

Overview

Purpose – why use Euclid to study geometry?

- to learn to reason deductively
- to instill in students an appreciation of an axiomatic system
- to learn the elements of geometry

Method – techniques most effective to accomplish the stated purposes

- minimal teacher input
- students present the proofs of all propositions, doing so as soon as possible without benefit of notes
- students who are not presenting ask questions, criticize, and give help when needed
- homework on almost all nights consists of studying/reviewing definitions, postulates, common notions, previous propositions, and of preparation of the latest proposition(s) for presentation the next day
- class participation, when presenting and when others are presenting, is the basis for most of a student's grade
- at critical points in the course, homework is assigned which consists of additional problems (usually requiring proofs) using the knowledge acquired to date

Difficulties inherent in the text

- use of Common Notions not explicitly stated (see Notes on Book I)
- Proposition 4 (see Notes on Book I)
- treatment of ratios (see Notes on Book V)

Organization of the material

- The following are omitted.

    - Book II
    - Book III, the last four propositions
    - Book IV, the last seven propositions
    - Book V, the last three propositions
    - Book VI, the last seven propositions

All the omitted propositions can be found, with their proofs, in Appendix A. None of the omitted propositions is referred to in later propositions (not the ones we cover, that is), so the flow of logic is not interrupted. (It happens that in the middle of Books I and III there are also a small number of propositions that are not used later in the text. They are mentioned by number in the Teacher's Notes in the back of the book.)

It should be noted that to complete Book III or Book IV it is necessary to cover Book II. Book II is very different from the other books, both in the types of relationships many of the propositions demonstrate (being the geometric representation of algebraic truths students have already been exposed to) and in the method of proof. While it is possible to prove the later propositions in Book II (up through Proposition II.8) using the earlier ones in Book II, Euclid makes the proofs independent and therefore repetitive. The advantage of this from the student's standpoint is that over the course of eight propositions he will develop a true understanding of what he is proving and why it needs to be proved at all. (The disadvantage is that if the he stops grappling with these issues then working through the proofs will seem somewhat tedious and Book II may feel quite long.) If you are fearful of not finishing all the books in one year, Book II can be omitted without any loss of material that would be covered in a modern geometry class. However, if you discover you are moving quickly through Book I, by all means include Book II in your course. You will then have the option of later covering some very interesting propositions (especially Proposition IV.10) which are impossible to cover without Book II.

- One proposition is added

The congruence propositions are among the most used in geometry. Euclid proves (to use modern abbreviations) SAS, SSS, AAS and ASA. Of course, in class you will want your students to discover a counterexample for SSA (which they can begin trying to find after Proposition I.4 and continue trying until they figure it out or it is assigned for homework) and AAA (which is trivial after Proposition I.32). Euclid does not have a proposition for the special case of two right triangles having one leg and the hypotenuse of one triangle equal to one leg and the hypotenuse of the other triangle, respectively. I have therefore included what I have called Proposition HL. Although it can be proved after Proposition I.26, I have put it in the exercise set following Proposition I.32.

- The concept of congruence is not used by Euclid

Euclid does not define congruence, but repeats the conclusion of his congruence propositions by stating that the two triangles are equal (meaning in area), and the sides and angles are equal respectively (whatever is appropriate depending on the hypothesis of the proposition).

Each Book in Brief

Book I: Triangles, basic constructions, parallel lines, parallelograms, the beginning of the theory of areas, the Pythagorean theorem

Book II: Geometric algebra, completion of the theory of areas, generalization of the Pythagorean theorem (geometric form of the law of cosines)

Book III: Circles, segments, sectors, tangents, angles

Book IV: Inscribed and circumscribed figures

Book V: Ratio and proportion

Book VI: Similar figures

How much time to expect to spend in each book

Expect to spend close to half the year in Book I. Learning proofs is difficult for some students and it is not unusual to spend a disproportionate amount of time in Book I. Book III (circles) goes a little faster, and Books IV and VI quite fast.

Getting started

The following is for teachers who have never encountered *The Elements* before and might want some suggestions as to how to proceed.

Definitions

Reading and discussing the definitions, postulates, and common notions will probably take one to two class periods. Today's students have almost certainly never read any mathematics before and they will need to be taught to read slowly and to think about the meaning of every word.

In addition, Euclid's way of expressing ideas is not a modern one. My own experience is that the greatest understanding comes when I do not explain things that seem difficult. As an example, consider Definition I.1: "a point is that which has no part." What is "a part" of something? What can it mean to "have no part?" As students grapple with the meaning of words they are forced to think slowly and precisely. And the definitions don't last forever. In a day or two the class will be on to more straightforward things with a good foundation.

During these early classes you will probably want to emphasize to your students that they are not learning the definitions, postulates, and common notions to pass a test, but that they will repeatedly be appealing to them all year long.

Working out the proofs

After a foundation of definitions, postulates, and common notions has been laid, the objective is to enable each student to work out each proof on his own in order to present it to the class. Thus it most important for each student to learn, before he starts to work on the propositions, what it is he is to prove. Even if he is unsuccessful at home in figuring out the proof, at the very least he should come to class each day being able to state without hesitation the enunciation of the proposition(s) due that day. He should also be able to identify what information is given and what statement is to be proved. My own experience is that the simplest and least time consuming way to accomplish this goal is to make routine two things, the total time for which takes about five to seven minutes of each class.

- First, at the beginning of each class, I have the students write down the enunciation of the day's proposition, and continue with the first two statements of the proof. That is, they write the enunciation, set up and letter the drawing, and conclude with a statement that begins, "I say that…." or "We have to prove that…" or some equivalent statement, just as Euclid does. This process can be timed to allow about 30 seconds longer than it takes the teacher.

- Second, for the last few minutes of each class, we "interpret" whatever propositions will be due the following day. Students learn how to "read with a pencil" and they figure out, as they read, what the proposition says. By the end of this time they should be able to put the proposition in at least one form for identifying a hypothesis and a conclusion (see Appendix B).

I use Proposition I.1 to teach the students how to "translate" Euclid. I have the students take notes on what I do and say. This is the first and last time I go to the board to present a proposition and my students know that the next day they must do with Proposition I.2 what they are watching me do with Proposition I.1. (See the notes on the propositions for more details.) I allow the students to use notes when making their presentations until they reach Proposition I.15. Proposition I.15 is a particularly easy proposition and a good one to start them working at the board without notes.

Supplemental work

Grappling with proofs in order to learn not only the basics of plane geometry but also how to think, is the whole point of studying Euclid. That being the case, one wants to be cautious about introducing supplemental material, lest it encourages the mentality so prevalent in modern geometry classes of, "I can do the problems but not the proofs."

Memory work

As far as memory work is concerned, I have found frequent but brief oral review ("Chris, please remind us what Prop 5 says") is very helpful and keeps the students on their toes. Also, I require certain of the most frequently used propositions to be known by number as well as its enunciation. This sounds daunting to the students at first, but when they have used Proposition 4, for example, (SAS) a certain number of times they begin to realize that they have memorized both its statement and its number without any extra effort. And certain groups of propositions, for example construction propositions 9 – 12, are related and it is easy to at least memorize that those four are constructions.

Table of Contents

Book I Definitions and Propositions 1 – 48   1
     Triangles, basic constructions, parallel lines, parallelograms,
     beginning of the theory of areas, Pythagorean theorem
     Exercises are on pages 6, 11, 16, 25, and 37

Book III Definitions and Propositions 1 – 34   41
     Circles, segments, sectors, tangents, angles
     Exercises are on pages 56, 67, and 72

Book IV Definitions and Propositions 1 – 9   73
     Inscribed and circumscribed figures
     Exercises are on page 79

Book V   Definitions and Propositions 1 – 22   81
     Ratios and proportions

Book VI Definitions and Propositions 1 – 13   99
     Similar figures
     Exercises are on page 110

Appendix A
     Book II   115
     Book III Props 35–37   127
     Book IV Props 10–16   131
     Book V Props 23–25   138
     Book VI Props 14–33   141

Appendix B   159
     On Logic and Deductive Reasoning

Teacher's Notes
     Book I   167
     Book II   172
     Book III   174
     Book IV   176
     Book V   177
     Book VI   182
     Notes on Selected Exercises   183

Book I

Definitions

1. A *point* is that which has no part.

2. A *line* is breadthless length.

3. The ends of a line are points.

4. A *straight line* is a line which lies evenly with the points on itself.

5. A *surface* is that which has length and breadth only.

6. The edges of a surface are lines.

7. A *plane surface* is a surface which lies evenly with the straight lines on itself.

8. A *plane angle* is the inclination to one another of two lines in a plane which meet one another and do not lie in a straight line.

9. And when the lines containing the angle are straight, the angle is called *rectilinear*.

10. When a straight line standing on a straight line makes the adjacent angles equal to one another, each of the equal angles is *right*, and the straight line standing on the other is called a *perpendicular* to that on which it stands.

11. An *obtuse angle* is an angle greater than a right angle.

12. An *acute angle* is an angle less than a right angle.

13. A *boundary* is that which is an extremity of anything.

14. A *figure* is that which is contained by any boundary or boundaries.

15. A *circle* is a plane figure contained by one line such that all the straight lines falling upon it from one point among those lying within the figure equal one another.

16. And the point is called the *center* of the circle.

17. A *diameter* of the circle is any straight line drawn through the center and terminated in both directions by the circumference of the circle, and such a straight line also bisects the circle.

18. A *semicircle* is the figure contained by the diameter and the circumference cut off by it. And the center of the semicircle is the same as that of the circle.

19. *Rectilinear figures* are those which are contained by straight lines, *trilateral* figures being those contained by three, *quadrilateral* those contained by four, and *multilateral* those contained by more than four straight lines.

20. Of trilateral figures, an *equilateral triangle* is that which has its three sides equal, an *isosceles triangle* that which has two of its sides alone equal, and a *scalene triangle* that which has its three sides unequal.

21. Further, of trilateral figures, a *right-angled triangle* is that which has a right angle, an *obtuse-angled triangle* that which has an obtuse angle, and an *acute-angled triangle* that which has its three angles acute.

22. Of quadrilateral figures, a *square* is that which is both equilateral and right-angled; an *oblong* that which is right-angled but not equilateral; a *rhombus* that which is equilateral but not right-angled; and a *rhomboid* that which has its opposite sides and angles equal to one another but is neither equilateral nor right-angled. And let quadrilaterals other than these be called *trapezia*.

23. *Parallel* straight lines are straight lines which, being in the same plane and being produced indefinitely in both directions, do not meet one another in either direction.

Postulates

1. To draw a straight line from any point to any point.

2. To produce a finite straight line continuously in a straight line.

3. To describe a circle with any center and radius.

4. That all right angles equal one another.

5. That, if a straight line falling on two straight lines makes the interior angles on the same side less than two right angles, the two straight lines, if produced indefinitely, meet on that side on which the angles are less than the two right angles.

Common Notions

1. Things which equal the same thing also equal one another.

2. If equals are added to equals, then the wholes are equal.

3. If equals are subtracted from equals, then the remainders are equal.

4. Things which coincide with one another equal one another.

5. The whole is greater than the part.

Book I Propositions

Proposition 1

To construct an equilateral triangle on a given finite straight line.

Let AB be the given finite straight line.

It is required to construct an equilateral triangle on the straight line AB.

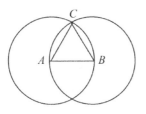

Describe the circle BCD with center A and radius AB. Again describe the circle ACE with center B and radius BA. Join the straight lines CA and CB from the point C at which the circles cut one another to the points A and B.   Post.3 Post.1

Now, since the point A is the center of the circle CDB, AC equals AB.   Def.15

Again, since the point B is the center of the circle CAE, BC equals BA.   Def.15

But AC was proved equal to AB, therefore each of the straight lines AC and BC equals AB.

And things which equal the same thing also equal one another, therefore AC also equals BC.   C.N.1

Therefore the three straight lines AC, AB, and BC equal one another.

Therefore the triangle ABC is equilateral, and it has been constructed on the given finite straight line AB.   Def.20

Q.E.F.

Proposition 2

To place a straight line equal to a given straight line with one end at a given point.

Let A be the given point, and BC the given straight line.

It is required to place a straight line equal to the given straight line BC with one end at the point A.

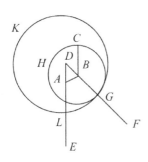

Join the straight line AB from the point A to the point B, and construct the equilateral triangle DAB on it.   Post. 1 I.1.

Produce the straight lines AE and BF in a straight line with DA and DB. Describe the circle CGH with center B and radius BC, and again, describe the circle GKL with center D and radius DG.   Post.2 Post.3

Since the point B is the center of the circle CGH, therefore BC equals BG. Again, since the point D is the center of the circle GKL, therefore DL equals DG.   I.Def.15

And in these *DA* equals *DB*, therefore the remainder *AL* equals the remainder *BG*.   C.N.3

But *BC* was also proved equal to *BG*, therefore each of the straight lines *AL* and *BC* equals *BG*. And things which equal the same thing also equal one another, therefore *AL* also equals *BC*.   C.N.1

Therefore the straight line *AL* equal to the given straight line *BC* has been placed with one end at the given point *A*.

Q.E.F.

Proposition 3

To cut off from the greater of two given unequal straight lines a straight line equal to the less.

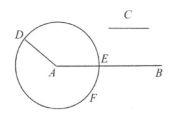

Let *AB* and *C* be the two given unequal straight lines, and let *AB* be the greater of them.

It is required to cut off from *AB* the greater a straight line equal to *C* the less.

Place *AD* at the point *A* equal to the straight line *C*, and describe the circle *DEF* with center *A* and radius *AD*.   I.2, Post. 3

Now, since the point *A* is the center of the circle *DEF*, therefore *AE* equals *AD*.   I.Def.15

But *C* also equals *AD*, therefore each of the straight lines *AE* and *C* equals *AD*, so that *AE* also equals *C*.   C.N.1

Therefore, given the two straight lines *AB* and *C*, *AE* has been cut off from *AB* the greater equal to *C* the less.

Q.E.F.

Proposition 4

If two triangles have two sides equal to two sides respectively, and have the angles contained by the equal straight lines equal, then they also have the base equal to the base, the triangle equals the triangle, and the remaining angles equal the remaining angles respectively, namely those opposite the equal sides.

Let *ABC* and *DEF* be two triangles having the two sides *AB* and *AC* equal to the two sides *DE* and *DF* respectively, namely *AB* equal to *DE* and *AC* equal to *DF*, and the angle *BAC* equal to the angle *EDF*.

I say that the base *BC* also equals the base *EF*, the triangle *ABC* equals the triangle *DEF*, and the remaining angles equal the remaining angles respectively, namely those opposite the equal sides, that is, the angle *ABC* equals the angle *DEF*, and the angle *ACB* equals the angle *DFE*.

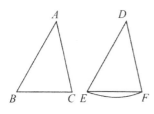

If the triangle ABC is superposed on the triangle DEF, and if the point A is placed on the point D and the straight line AB on DE, then the point B also coincides with E, because AB equals DE.

Again, AB coinciding with DE, the straight line AC also coincides with DF, because the angle BAC equals the angle EDF. Hence the point C also coincides with the point F, because AC again equals DF.

But B also coincides with E, hence the base BC coincides with the base EF and equals it.
[For if, when B coincides with E and C with F, the base BC does not coincide with the base EF, two straight lines will enclose a space: which is impossible.]
Therefore the base BC will coincide with EF and will be equal to it.   C.N.4
Thus the whole triangle ABC coincides with the whole triangle DEF and equals it. And the remaining angles also coincide with the remaining angles and equal them, the angle ABC equals the angle DEF, and the angle ACB equals the angle DFE.
Therefore *if two triangles have two sides equal to two sides respectively, and have the angles contained by the equal straight lines equal, then they also have the base equal to the base, the triangle equals the triangle, and the remaining angles equal the remaining angles respectively, namely those opposite the equal sides.*

Q.E.D.

Proposition 5

In isosceles triangles the angles at the base equal one another, and, if the equal straight lines are produced further, then the angles under the base equal one another.

Let ABC be an isosceles triangle having the side AB equal to the side AC, and let the straight lines BD and CE be produced further in a straight line with AB and AC.   I.Def.20
                                                                                                Post.2

I say that the angle ABC equals the angle ACB, and the angle CBD equals the angle BCE.

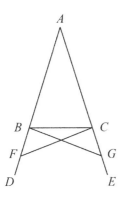

Take an arbitrary point F on BD. Cut off AG from AE the greater equal to AF the less, and join the straight lines FC and GB.   I.3
                                                                                                Post.1

Since AF equals AG, and AB equals AC, therefore the two sides FA and AC equal the two sides GA and AB, respectively, and they contain a common angle, the angle FAG.

Therefore the base FC equals the base GB, the triangle AFC equals the triangle AGB, and the remaining angles equal the remaining angles respectively, namely those opposite the equal sides, that is, the angle ACF equals the angle ABG, and the angle AFC equals the angle AGB.   I.4

Since the whole AF equals the whole AG, and in these AB equals AC, therefore the remainder BF equals the remainder CG. — C.N.3

But FC was also proved equal to GB, therefore the two sides BF and FC equal the two sides CG and GB respectively, and the angle BFC equals the angle CGB, while the base BC is common to them. Therefore the triangle BFC also equals the triangle CGB, and the remaining angles equal the remaining angles respectively, namely those opposite the equal sides. Therefore the angle FBC equals the angle GCB, and the angle BCF equals the angle CBG. — I.4

Accordingly, since the whole angle ABG was proved equal to the angle ACF, and in these the angle CBG equals the angle BCF, the remaining angle ABC equals the remaining angle ACB, and they are at the base of the triangle ABC. But the angle FBC was also proved equal to the angle GCB, and they are under the base. — C.N.3

Therefore *in isosceles triangles the angles at the base equal one another, and, if the equal straight lines are produced further, then the angles under the base equal one another.*

Q.E.D.

Exercises on Proofs

1. Proposition: An equilateral triangle is also equiangular. Prove this proposition in the manner of Euclid. That is, make a drawing, label it, restate what you want to prove in terms of your drawing, then prove it.

2. Draw line AB about 3" long, then follow the directions below. For parts (a) through (e), give reasons why you can do what you are doing.

   a. Draw equilateral triangle ABC.

   b. Pick any point D on AB (not too near the midpoint).

   c. Place E on CB so that CE = BD.

   d. Place F on AC so that AF = BD.

   e. Draw DE, EF, and FD.

   f. Prove that triangle DEF is equilateral.

3. Given circle ABC with center D, prove that angle ACB is equal to the sum of angle CAD and angle CBD.

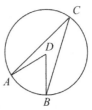

Proposition 6

If in a triangle two angles equal one another, then the sides opposite the equal angles also equal one another.

Let ABC be a triangle having the angle ABC equal to the angle ACB.

I say that the side AB also equals the side AC.

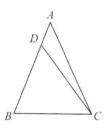

If AB does not equal AC, then one of them is greater.	C.N
Let AB be greater. Cut off DB from AB the greater equal to AC the less, and join DC.	I.3 Post.1
Since DB equals AC, and BC is common, therefore the two sides DB and BC equal the two sides AC and CB respectively, and the angle DBC equals the angle ACB. Therefore the base DC equals the base AB, and the triangle DBC equals the triangle ACB, the less equals the greater, which is absurd.	I.4 C.N.5

Therefore AB is not unequal to AC, it therefore equals it.

Therefore *if in a triangle two angles equal one another, then the sides opposite the equal angles also equal one another.*

Q.E.D.

Proposition 7

Given two straight lines constructed from the ends of a straight line and meeting in a point, there cannot be constructed from the ends of the same straight line, and on the same side of it, two other straight lines meeting in another point and equal to the former two respectively, namely each equal to that from the same end.

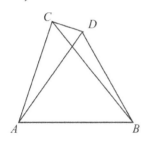

If possible, given two straight lines AC and CB constructed on the straight line AB and meeting at the point C, let two other straight lines AD and DB be constructed on the same straight line AB, on the same side of it, meeting in another point D and equal to the former two respectively, namely each equal to that from the same end, so that AC equals AD which has the same end A, and CB equals DB which has the same end B.

Join CD.	Post.1
Since AC equals AD, therefore the angle ACD equals the angle ADC. Therefore the angle ADC is greater than the angle DCB. Therefore the angle CDB is much greater than the angle DCB.	I.5 C.N.5 C.N.
Again, since CB equals DB, therefore the angle CDB also equals the angle DCB. But it was also proved much greater than it, which is impossible.	I.5, C.N.

Therefore *given two straight lines constructed from the ends of a straight line and meeting in a point, there cannot be constructed from the ends of the same straight line, and on the same side of it, two other straight lines meeting in another point and equal to the former two respectively, namely each equal to that from the same end.*

Q.E.D.

Proposition 8

If two triangles have the two sides equal to two sides respectively, and also have the base equal to the base, then they also have the angles equal which are contained by the equal straight lines.

Let ABC and DEF be two triangles having the two sides AB and AC equal to the two sides DE and DF respectively, namely AB equal to DE and AC equal to DF, and let them have the base BC equal to the base EF.

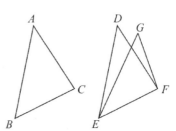

I say that the angle BAC also equals the angle EDF.

For, if the triangle ABC is applied to the triangle DEF, and if the point B is placed on the point E and the straight line BC on EF, then the point C also coincides with F, because BC equals EF.

Then, BC coinciding with EF, therefore BA and AC also coincide with ED and DF, for, if the base BC coincides with the base EF, and the sides BA and AC do not coincide with ED and DF but fall beside them as EG and GF, then given two straight lines constructed on a straight line and meeting in a point, there will have been constructed on the same straight line and on the same side of it, two other straight lines meeting in another point and equal to the former two respectively, namely each to that which has the same end with it.

But they cannot be so constructed.   I.7

Therefore it is not possible that, if the base BC is applied to the base EF, the sides BA and AC do not coincide with ED and DF. Therefore they coincide, so that the angle   C.N.4
BAC coincides with the angle EDF, and equals it.

Therefore *if two triangles have the two sides equal to two sides respectively, and also have the base equal to the base, then they also have the angles equal which are contained by the equal straight lines.*

Q.E.D.

Review Questions

Make sure at this point that you can answer the following questions without hesitation.

1. What does it mean to say two triangles are equal?
2. What is an indirect proof? Of the propositions you have proven thus far, which was/were proved indirectly?
3. State the postulates and common notions of Euclid.
4. State propositions I.1 through I.8

Proposition 9

To bisect a given rectilinear angle.
Let the angle *BAC* be the given rectilinear angle.
It is required to bisect it.

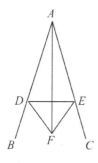

Take an arbitrary point *D* on *AB*. Cut off *AE* from *AC* equal to *AD*, and join *DE*. Construct the equilateral triangle *DEF* on *DE*, and join *AF*. I.3, Post.1, I.1

I say that the angle *BAC* is bisected by the straight line *AF*.

Since *AD* equals *AE*, and *AF* is common, therefore the two sides *AD* and *AF* equal the two sides *EA* and *AF* respectively.

And the base *DF* equals the base *EF*, therefore the angle *DAF* equals the angle *EAF*. I.Def.20, I.8

Therefore the given rectilinear angle *BAC* is bisected by the straight line *AF*.

Q.E.F.

Proposition 10

To bisect a given finite straight line.
Let *AB* be the given finite straight line.

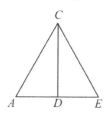

It is required to bisect the finite straight line *AB*.
Construct the equilateral triangle *ABC* on it, and bisect the angle *ACB* by the straight line *CD*. I.1, I.9

I say that the straight line *AB* is bisected at the point *D*.

Since *CA* equals *CB*, and *CD* is common, therefore the two sides *CA* and *CD* equal the two sides *CB* and *CD* respectively, and the angle *ACD* equals the angle *BCD*, therefore the base *AD* equals the base *BD*. I.Def.20, I.4

Therefore the given straight line *AB* is bisected at *D*.

Q.E.F.

Proposition 11

To draw a straight line at right angles to a given straight line from a given point on it.

Let AB be the given straight line, and C the given point on it.
It is required to draw a straight line at right angles to the straight line AB from the point C.

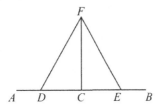

Take an arbitrary point D on AC. Make CE equal to CD. Construct the equilateral triangle FDE on DE, and join CF.

I.3, I.1
Post.1

I say that the straight line CF has been drawn at right angles to the given straight line AB from C the given point on it.

Since CD equals CE, and CF is common, therefore the two sides CD and CF equal the two sides CE and CF respectively, and the base DF equals the base EF. Therefore the angle DCF equals the angle ECF, and they are adjacent angles.

I.Def.20
I.8

But, when a straight line standing on a straight line makes the adjacent angles equal to one another, each of the equal angles is right, therefore each of the angles DCF and FCE is right.

I.Def.10

Therefore the straight line CF has been drawn at right angles to the given straight line AB from the given point C on it.

Q.E.F.

Proposition 12

To draw a straight line perpendicular to a given infinite straight line from a given point not on it.

Let AB be the given infinite straight line, and C the given point which is not on it.
It is required to draw a straight line perpendicular to the given infinite straight line AB from the given point C which is not on it.

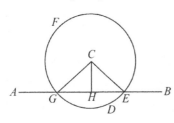

Take an arbitrary point D on the other side of the straight line AB, and describe the circle EFG with center C and radius CD. Bisect the straight line EG at H, and join the straight lines CG, CH, and CE.

Post.3
I.10
Post.1

I say that CH has been drawn perpendicular to the given infinite straight line AB from the given point C which is not on it.

Since GH equals HE, and HC is common, therefore the two sides GH and HC equal the two sides EH and HC respectively, and the base CG equals the base CE. Therefore the angle CHG equals the angle EHC, and they are adjacent angles.

I.Def.15
I.8

But, when a straight line standing on a straight line makes the adjacent angles equal to one another, each of the equal angles is right, and the straight line standing on the other is called a perpendicular to that on which it stands.   I.Def.10

Therefore CH has been drawn perpendicular to the given infinite straight line AB from the given point C which is not on it.

Q.E.F.

Construction Exercises

1. In the middle of a full sized sheet of plain paper, using a straightedge, draw an acute, scalene triangle (with sides in the neighborhood of three inches long). Using compass and straightedge, carefully bisect each side of your triangle to find the midpoints of the sides. Connect each midpoint to the vertex opposite it. Do the lines appear to intersect in a single point? Try the same thing with other triangles of different shapes.

2. In the middle of a full sized sheet of plain paper, using a straightedge, draw an obtuse, scalene triangle (with sides in the neighborhood of three inches long). Using compass and straightedge, carefully bisect each angle of your triangle and extend the bisectors to the opposite sides of the triangle. Do the bisectors appear to intersect in one point?

3. In the middle of a full sized sheet of plain paper, using a straightedge, draw a triangle with sides in the neighborhood of three inches long. Using compass and straightedge, from each vertex carefully construct a line perpendicular to the side opposite. (If you draw an obtuse triangle you will need to extend the side next to the obtuse angle in order to construct the perpendicular.) Do the perpendiculars appear to intersect in one point?

4. On a full sized sheet of paper, with a straightedge, draw a triangle about the size of any one of your other three triangles. Using a compass and straightedge, carefully construct perpendicular bisectors of each side of your triangle. Extend the bisectors until they meet one another. Again, do the lines appear to intersect in a single point?

5. Construct a triangle in which one angle is one-half of a right angle and another angle is one-quarter of a right angle.

6. On a separate piece of paper, draw a horizontal line about three inches long and call it AB. Using AB as base, construct a triangle that is isosceles but not equilateral. Prove your result. That is, prove that the triangle you construct is (a) isosceles, and (b) not equilateral.

Proposition 13

If a straight line stands on a straight line, then it makes either two right angles or angles whose sum equals two right angles.

Let any straight line AB standing on the straight line CD make the angles CBA and ABD.
I say that either the angles CBA and ABD are two right angles or their sum equals two right angles.

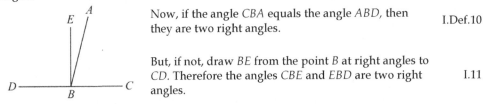

Now, if the angle CBA equals the angle ABD, then they are two right angles.   I.Def.10

But, if not, draw BE from the point B at right angles to CD. Therefore the angles CBE and EBD are two right angles.   I.11

Since the angle CBE equals the sum of the two angles CBA and ABE, add the angle EBD to each, therefore the sum of the angles CBE and EBD equals the sum of the three angles CBA, ABE, and EBD.   C.N.2

Again, since the angle DBA equals the sum of the two angles DBE and EBA, add the angle ABC to each, therefore the sum of the angles DBA and ABC equals the sum of the three angles DBE, EBA, and ABC.   C.N.2

But the sum of the angles CBE and EBD was also proved equal to the sum of the same three angles, and things which equal the same thing also equal one another, therefore the sum of the angles CBE and EBD also equals the sum of the angles DBA and ABC. But the angles CBE and EBD are two right angles, therefore the sum of the angles DBA and ABC also equals two right angles.   C.N.1

Therefore *if a straight line stands on a straight line, then it makes either two right angles or angles whose sum equals two right angles.*

Q.E.D.

Proposition 14

If with any straight line, and at a point on it, two straight lines not lying on the same side make the sum of the adjacent angles equal to two right angles, then the two straight lines are in a straight line with one another.

With any straight line AB, and at the point B on it, let the two straight lines BC and BD not lying on the same side make the sum of the adjacent angles ABC and ABD equal to two right angles.
I say that BD is in a straight line with CB.

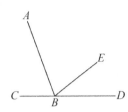

If *BD* is not in a straight line with *BC*, then produce *BE* in a straight line with *CB*. — Post.2

Since the straight line *AB* stands on the straight line *CBE*, therefore the sum of the angles *ABC* and *ABE* equals two right angles. But the sum of the angles *ABC* and *ABD* also equals two right angles, therefore the sum of the angles *CBA* and *ABE* equals the sum of the angles *CBA* and *ABD*. — I.13, Post.4, C.N.1

Subtract the angle *CBA* from each. Then the remaining angle *ABE* equals the remaining angle *ABD*, the less equals the greater, which is impossible. Therefore *BE* is not in a straight line with *CB*. — C.N.3

Similarly we can prove that neither is any other straight line except *BD*. Therefore *CB* is in a straight line with *BD*.

Therefore *if with any straight line, and at a point on it, two straight lines not lying on the same side make the sum of the adjacent angles equal to two right angles, then the two straight lines are in a straight line with one another.*

Q.E.D.

Proposition 15

If two straight lines cut one another, then they make the vertical angles equal to one another.
Let the straight lines *AB* and *CD* cut one another at the point *E*.
I say that the angle *CEA* equals the angle *DEB*, and the angle *BEC* equals the angle *AED*.

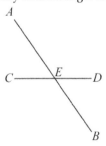

Since the straight line *AE* stands on the straight line *CD* making the angles *CEA* and *AED*, therefore the sum of the angles *CEA* and *AED* equals two right angles. — I.13

Again, since the straight line *DE* stands on the straight line *AB* making the angles *AED* and *DEB*, therefore the sum of the angles *AED* and *DEB* equals two right angles. — I.13

But the sum of the angles *CEA* and *AED* was also proved equal to two right angles, therefore the sum of the angles *CEA* and *AED* equals the sum of the angles *AED* and *DEB*. — Post.4, C.N.1

Subtract the angle *AED* from each. Then the remaining angle *CEA* equals the remaining angle *DEB*. — C.N.3

Similarly it can be proved that the angles *BEC* and *AED* are also equal.

Therefore *if two straight lines cut one another, then they make the vertical angles equal to one another.*

Q.E.D.

Proposition 16

In any triangle, if one of the sides is produced, then the exterior angle is greater than either of the interior and opposite angles.

Let *ABC* be a triangle, and let one side of it *BC* be produced to *D*.
I say that the exterior angle *ACD* is greater than either of the interior and opposite angles *CBA* and *BAC*.

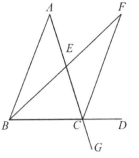

Bisect *AC* at *E*. Join *BE*, and produce it in a straight line to *F*.

I.10
Post.1
Post.2

Make *EF* equal to *BE*, join *FC*, and draw *AC* through to *G*.

I.3
Post.1
Post.2

Since *AE* equals *EC*, and *BE* equals *EF*, therefore the two sides *AE* and *EB* equal the two sides *CE* and *EF* respectively, and the angle *AEB* equals the angle *FEC*, for they are vertical angles.

I.15

Therefore the base *AB* equals the base *FC*, the triangle *ABE* equals the triangle *CFE*, and the remaining angles equal the remaining angles respectively, namely those opposite the equal sides. Therefore the angle *BAE* equals the angle *ECF*.

I.4

But the angle *ECD* is greater than the angle *ECF*, therefore the angle *ACD* is greater than the angle *BAE*.

C.N.5

Similarly, if *BC* is bisected, then the angle *BCG*, that is, the angle *ACD*, can also be proved to be greater than the angle *ABC*.

I.15

Therefore *in any triangle, if one of the sides is produced, then the exterior angle is greater than either of the interior and opposite angles.*

Q.E.D.

Proposition 17

In any triangle the sum of any two angles is less than two right angles.

Let *ABC* be a triangle.
I say that the sum of any two angles of the triangle *ABC* is less than two right angles.

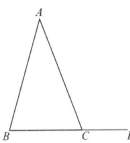

Produce *BC* to *D*.

Post.2

Since the angle *ACD* is an exterior angle of the triangle *ABC*, therefore it is greater than the interior and opposite angle *ABC*. Add the angle *ACB* to each. Then the sum of the angles *ACD* and *ACB* is greater than the sum of the angles *ABC* and *BCA*.

I.16
C.N.

But the sum of the angles *ACD* and *ACB* is equal to two right angles. Therefore the sum of the angles *ABC* and *BCA* is less than two right angles.

I.13

Similarly we can prove that the sum of the angles *BAC* and *ACB* is also less than two right angles, and so the sum of the angles *CAB* and *ABC* as well.
Therefore *in any triangle the sum of any two angles is less than two right angles.*

Q.E.D.

Proposition 18

In any triangle the angle opposite the greater side is greater.

Let ABC be a triangle having the side AC greater than AB.
I say that the angle ABC is also greater than the angle BCA.

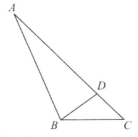

Since AC is greater than AB, make AD equal to AB, and join BD.   I.3, Post.1

Since the angle ADB is an exterior angle of the triangle BCD, therefore it is greater than the interior and opposite angle DCB.   I.16

But the angle ADB equals the angle ABD, since the side AB equals AD, therefore the angle ABD is also greater than the angle ACB. Therefore the angle ABC is much greater than the angle ACB.   I.5

Therefore *in any triangle the angle opposite the greater side is greater.*

Q.E.D.

Proposition 19

In any triangle the side opposite the greater angle is greater.

Let ABC be a triangle having the angle ABC greater than the angle BCA.
I say that the side AC is greater than the side AB.

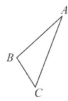

If not, either AC equals AB or it is less than it.

Now AC does not equal AB, for then the angle ABC would equal the angle ACB, but it does not. Therefore AC does not equal AB.   I.5

Neither is AC less than AB, for then the angle ABC would be less than the angle ACB, but it is not. Therefore AC is not less than AB.   I.18

And it was proved that it is not equal either. Therefore AC is greater than AB.

Therefore *in any triangle the side opposite the greater angle is greater.*

Q.E.D.

Proposition 20

In any triangle the sum of any two sides is greater than the remaining one.

Let ABC be a triangle.
I say that in the triangle ABC the sum of any two sides is greater than the remaining one, that is, the sum of BA and AC is greater than BC, the sum of AB and BC is greater than AC, and the sum of BC and CA is greater than AB.

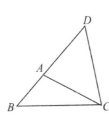

Draw *BA* through to the point *D*, and make *DA* equal to *CA*. Join *DC*.

Since *DA* equals *AC*, therefore the angle *ADC* also equals the angle *ACD*. Therefore the angle *BCD* is greater than the angle *ADC*.

Since *DCB* is a triangle having the angle *BCD* greater than the angle *BDC*, and the side opposite the greater angle is greater, therefore *DB* is greater than *BC*.

Post.2, I.3
Post.1
I.5
C.N.5
I.19

But *DA* equals *AC*, therefore the sum of *BA* and *AC* is greater than *BC*.
Similarly we can prove that the sum of *AB* and *BC* is also greater than *CA*, and the sum of *BC* and *CA* is greater than *AB*.

Therefore *in any triangle the sum of any two sides is greater than the remaining one.*

Q.E.D.

Exercises on Propositions I.1 – I.20

1. Given: △*BAD* and △*BCD* are isosceles
   Prove: ∠*BEA* is a right angle

   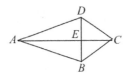

2. Given: *BD* bisects ∠*ABC*
   Prove: *AB* > *AD*

   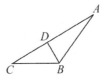

3. Given: ∠*EAB* > ∠*EBA*; ∠*ECD* > ∠*EDC*
   Prove: *BD* > *AC*

   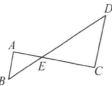

4. Given: △*ABC* is equilateral
   Prove: *AD* < *BC*

5. Given: *AB* = *AC*; ∠*DBC* = ∠*DCB*
   Prove: △*ADB* = △*ADC*

   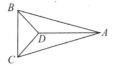

6. Given: Quadrilateral *ABCD*
   Prove: The sum of *AB*, *BC*, and *CD* > *AD*

   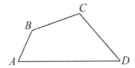

7. You know that the sum of any two sides of a triangle is greater than the third side. (Which proposition asserts that?) Do you think it is it true that the sum of any two angles of a triangle is greater than the third angle? Try to justify your answer. Can you?

8. Given three points A, B, and C, having the property that the line AB together with the line BC equals the line AC, prove that A, B, and C must lie in a straight line.

9. For each of the following statements, decide whether or not it is true. If true, prove the statement. If not true, give an example to show why not.

   a. An exterior angle of a triangle is greater than any interior angle.
   b. If two sides of a triangle are unequal, then the angles opposite them are unequal.
   c. The smallest angle of a triangle must be opposite its shortest side.

Proposition 21

If from the ends of one of the sides of a triangle two straight lines are constructed meeting within the triangle, then the sum of the straight lines so constructed is less than the sum of the remaining two sides of the triangle, but the constructed straight lines contain a greater angle than the angle contained by the remaining two sides.

From the ends B and C of one of the sides BC of the triangle ABC, let the two straight lines BD and DC be constructed meeting within the triangle.

I say that the sum of BD and DC is less than the sum of the remaining two sides of the triangle BA and AC, but BD and DC contain an angle BDC greater than the angle BAC.

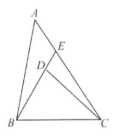

Draw BD through to E.	Post.2
Since in any triangle the sum of two sides is greater than the remaining one, therefore, in the triangle ABE, the sum of the two sides AB and AE is greater than BE.	I.20
Add EC to each. Then the sum of BA and AC is greater than the sum of BE and EC.	C.N.
Again, since, in the triangle CED, the sum of the two sides CE and ED is greater than CD, add DB to each, therefore the sum of CE and EB is greater than the sum of CD and DB.	I.20 C.N.
But the sum of BA and AC was proved greater than the sum of BE and EC, therefore the sum of BA and AC is much greater than the sum of BD and DC.	C.N.
Again, since in any triangle the exterior angle is greater than the interior and opposite angle, therefore, in the triangle CDE, the exterior angle BDC is greater than the angle CED.	I.16
For the same reason, moreover, in the triangle ABE the exterior angle CEB is greater than the angle BAC. But the angle BDC was proved greater than the angle CEB, therefore the angle BDC is much greater than the angle BAC.	I.16 C.N.

Therefore *if from the ends of one of the sides of a triangle two straight lines are constructed meeting within the triangle, then the sum of the straight lines so constructed is less than the sum of the remaining two sides of the triangle, but the constructed straight lines contain a greater angle than the angle contained by the remaining two sides.*

Q.E.D.

Proposition 22

To construct a triangle out of three straight lines which equal three given straight lines: thus it is necessary that the sum of any two of the straight lines should be greater than the remaining one.

Let the three given straight lines be A, B, and C, and let the sum of any two of these be greater than the remaining one, namely, A plus B greater than C, A plus C greater than B, and B plus C greater than A.
It is required to construct a triangle out of straight lines equal to A, B, and C.

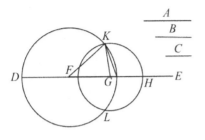

Set out a straight line DE, terminated at D but of infinite length in the direction of E. Make DF equal to A, FG equal to B, and GH equal to C.	Post.2 I.3
Describe the circle DKL with center F and radius FD. Again, describe the circle KLH with center G and radius GH. Join KF and KG.	Post.3 Post.1

I say that the triangle KFG has been constructed out of three straight lines equal to A, B, and C.

Since the point F is the center of the circle DKL, therefore FD equals FK. But FD equals A, therefore KF also equals A.	I.Def.16 C.N.1
Again, since the point G is the center of the circle LKH, therefore GH equals GK. But GH equals C, therefore KG also equals C.	I.Def.16 C.N.1

And FG also equals B, therefore the three straight lines KF, FG, and GK equal the three straight lines A, B, and C.

Therefore out of the three straight lines KF, FG, and GK, which equal the three given straight lines A, B, and C, the triangle KFG has been constructed.

Q.E.F

Proposition 23

To construct a rectilinear angle equal to a given rectilinear angle on a given straight line and at a point on it.
Let the angle DCE be the given rectilinear angle, AB the given straight line, and A the point on it.
It is required to construct a rectilinear angle equal to the given rectilinear angle DCE on the given straight line AB and at the point A on it.

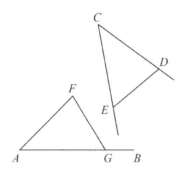

Take the points D and E at random on the straight lines CD and CE respectively, and join DE. Out of three straight lines which equal the three straight lines CD, DE, and CE construct the triangle AFG in such a way that CD equals AF, CE equals AG, and DE equals FG.

Post.1
I.22

Since the two sides DC and CE equal the two sides FA and AG respectively, and the base DE equals the base FG, therefore the angle DCE equals the angle FAG.

I.8

Therefore on the given straight line AB, and at the point A on it, the rectilinear angle FAG has been constructed equal to the given rectilinear angle DCE.

Q.E.F.

Proposition 24

If two triangles have two sides equal to two sides respectively, but have one of the angles contained by the equal straight lines greater than the other, then they also have the base greater than the base.
Let ABC and DEF be two triangles having the two sides AB and AC equal to the two sides DE and DF respectively, so that AB equals DE, and AC equals DF, and let the angle at A be greater than the angle at D.
I say that the base BC is greater than the base EF.

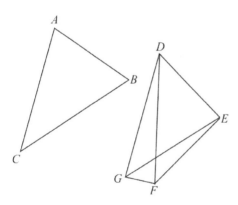

Since the angle BAC is greater than the angle EDF, construct the angle EDG equal to the angle BAC at the point D on the straight line DE. Make DG equal to either of the two straight lines AC or DF. Join EG and FG.

I.23
I.3
Post.1
I.4

Since AB equals DE, and AC equals DG, the two sides BA and AC equal the two sides ED and DG, respectively, and the angle BAC equals the angle EDG, therefore the base BC equals the base EG.

I.4

Again, since DF equals DG, therefore the angle DGF equals the angle DFG. Therefore the angle DFG is greater than the angle EGF.
Therefore the angle EFG is much greater than the angle EGF.

I.5

Since EFG is a triangle having the angle EFG greater than the angle EGF, and side opposite the greater angle is greater, therefore the side EG is also greater than EF.   I.19

But EG equals BC, therefore BC is also greater than EF.

Therefore *if two triangles have two sides equal to two sides respectively, but have one of the angles contained by the equal straight lines greater than the other, then they also have the base greater than the base.*

Q.E.D

Proposition 25

If two triangles have two sides equal to two sides respectively, but have the base greater than the base, then they also have the one of the angles contained by the equal straight lines greater than the other.

Let ABC and DEF be two triangles having two sides AB and AC equal to two sides DE and DF respectively, namely AB to DE, and AC to DF, and let the base BC be greater than the base EF. I say that the angle BAC is also greater than the angle EDF.

If not, it either equals it or is less.

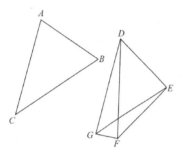

Now the angle BAC does not equal the angle EDF, for then the base BC would equal the base EF, but it is not. Therefore the angle BAC does not equal the angle EDF.   I.4

Neither is the angle BAC less than the angle EDF, for then the base BC would be less than the base EF, but it is not. Therefore the angle BAC is not less than the angle EDF.   I.24

But it was proved that it is not equal either. Therefore the angle BAC is greater than the angle EDF.

Therefore *if two triangles have two sides equal to two sides respectively, but have the base greater than the base, then they also have the one of the angles contained by the equal straight lines greater than the other.*

Q.E.D.

Proposition 26

If two triangles have two angles equal to two angles respectively, and one side equal to one side, namely, either the side adjoining the equal angles, or that opposite one of the equal angles, then the remaining sides equal the remaining sides and the remaining angle equals the remaining angle.

Let ABC and DEF be two triangles having the two angles ABC and BCA equal to the two angles DEF and EFD respectively, namely the angle ABC to the angle DEF, and the angle BCA to the angle EFD, and let them also have one side equal to one side, first that adjoining the equal angles, namely BC equal to EF.

I say that the remaining sides equal the remaining sides respectively, namely AB equals DE and AC equals DF, and the remaining angle equals the remaining angle, namely the angle BAC equals the angle EDF.

If *AB* does not equal *DE*, then one of them is greater.

Let *AB* be greater. Make *BG* equal to *DE*, and join *GC*.   I.3, Post 1

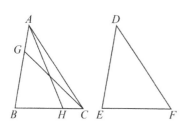

Since *BG* equals *DE*, and *BC* equals *EF*, the two sides *GB* and *BC* equal the two sides *DE* and *EF* respectively, and the angle *GBC* equals the angle *DEF*, therefore the base *GC* equals the base *DF*, the triangle *GBC* equals the triangle *DEF*, and the remaining angles equal the remaining angles, namely those opposite the equal sides. Therefore the angle *GCB* equals the angle *DFE*.   I.4

But the angle *DFE* equals the angle *ACB* by hypothesis. Therefore the angle *BCG* equals the angle *BCA*, the less equals the greater, which is impossible.   C.N.1

Therefore *AB* is not unequal to *DE*, and therefore equals it.

But *BC* also equals *EF*. Therefore the two sides *AB* and *BC* equal the two sides *DE* and *EF* respectively, and the angle *ABC* equals the angle *DEF*. Therefore the base *AC* equals the base *DF*, and the remaining angle *BAC* equals the remaining angle *EDF*.   I.4

Next, let sides opposite equal angles be equal, as *AB* equals *DE*.
I say again that the remaining sides equal the remaining sides, namely *AC* equals *DF* and *BC* equals *EF*, and further the remaining angle *BAC* equals the remaining angle *EDF*. If *BC* is unequal to *EF*, then one of them is greater.

Let *BC* be greater, if possible. Make *BH* equal to *EF*, and join *AH*.

Since *BH* equals *EF*, and *AB* equals *DE*, the two sides *AB* and *BH* equal the two sides *DE* and *EF* respectively, and they contain equal angles, therefore the base *AH* equals the base *DF*, the triangle *ABH* equals the triangle *DEF*, and the remaining angles equal the remaining angles, namely those opposite the equal sides. Therefore the angle *BHA* equals the angle *EFD*.   I.4

But the angle *EFD* equals the angle *BCA*, therefore, in the triangle *AHC*, the exterior angle *BHA* equals the interior and opposite angle *BCA*, which is impossible.   C.N. 1, I.16

Therefore *BC* is not unequal to *EF*, and therefore equals it.

But *AB* also equals *DE*. Therefore the two sides *AB* and *BC* equal the two sides *DE* and *EF* respectively, and they contain equal angles. Therefore the base *AC* equals the base *DF*, the triangle *ABC* equals the triangle *DEF*, and the remaining angle *BAC* equals the remaining angle *EDF*.   I.4

Therefore if two triangles have two angles equal to two angles respectively, and one side equal to one side, namely, either the side adjoining the equal angles, or that opposite one of the equal angles, then the remaining sides equal the remaining sides and the remaining angle equals the remaining angle.

Q.E.D.

Proposition 27

If a straight line falling on two straight lines makes the alternate angles equal to one another, then the straight lines are parallel to one another.

Let the straight line EF falling on the two straight lines AB and CD make the alternate angles AEF and EFD equal to one another.

I say that AB is parallel to CD.

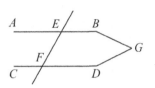

If not, AB and CD when produced meet either in the direction of B and D or towards A and C. Let them be produced and meet, in the direction of B and D, at G.

Then, in the triangle GEF, the exterior angle AEF equals the interior and opposite angle EFG, which is impossible. I.16

Therefore AB and CD when produced do not meet in the direction of B and D.
Similarly it can be proved that neither do they meet towards A and C.
But straight lines which do not meet in either direction are parallel. Therefore AB is parallel to CD. I.Def.23

Therefore *if a straight line falling on two straight lines makes the alternate angles equal to one another, then the straight lines are parallel to one another.*

Q.E.D.

Proposition 28

If a straight line falling on two straight lines makes the exterior angle equal to the interior and opposite angle on the same side, or the sum of the interior angles on the same side equal to two right angles, then the straight lines are parallel to one another.

Let the straight line EF falling on the two straight lines AB and CD make the exterior angle EGB equal to the interior and opposite angle GHD, or the sum of the interior angles on the same side, namely BGH and GHD, equal to two right angles.

I say that AB is parallel to CD.

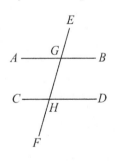

Since the angle EGB equals the angle GHD, and the angle EGB equals the angle AGH, therefore the angle AGH equals the angle GHD. And they are alternate, therefore AB is parallel to CD. I.15 C.N.1 I.27

Next, since the sum of the angles BGH and GHD equals two right angles, and the sum of the angles AGH and BGH also equals two right angles, therefore the sum of the angles AGH and BGH equals the sum of the angles BGH and GHD. I.13 C.N.1 Post.4

Subtract the angle BGH from each. Therefore the remaining angle AGH equals the remaining angle GHD. And they are alternate, therefore AB is parallel to CD. C.N.3 I.27

Therefore *if a straight line falling on two straight lines makes the exterior angle equal to the interior and opposite angle on the same side, or the sum of the interior angles on the same side equal to two right angles, then the straight lines are parallel to one another.*

Q.E.D.

Proposition 29

A straight line falling on parallel straight lines makes the alternate angles equal to one another, the exterior angle equal to the interior and opposite angle, and the sum of the interior angles on the same side equal to two right angles.

Let the straight line *EF* fall on the parallel straight lines *AB* and *CD*.

I say that it makes the alternate angles *AGH* and *GHD* equal, the exterior angle *EGB* equal to the interior and opposite angle *GHD*, and the sum of the interior angles on the same side, namely *BGH* and *GHD*, equal to two right angles.

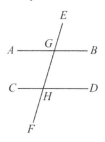

If the angle *AGH* does not equal the angle *GHD*, then one of them is greater. Let the angle *AGH* be greater.

Add the angle *BGH* to each. Therefore the sum of the angles *AGH* and *BGH* is greater than the sum of the angles *BGH* and *GHD*.

But sum of the angles *AGH* and *BGH* equals two right angles. Therefore the sum of the angles *BGH* and *GHD* is less than two right angles.   I.13

But straight lines produced indefinitely from angles less than two right angles meet. Therefore *AB* and *CD*, if produced indefinitely, will meet. But they do not meet, because they are by hypothesis parallel.   Post.5

Therefore the angle *AGH* is not unequal to the angle *GHD*, and therefore equals it.

Again, the angle *AGH* equals the angle *EGB*. Therefore the angle *EGB* also equals the angle *GHD*.   I.15 C.N.1

Add the angle *BGH* to each. Therefore the sum of the angles *EGB* and *BGH* equals the sum of the angles *BGH* and *GHD*.   C.N.2

But the sum of the angles *EGB* and *BGH* equals two right angles. Therefore the sum of the angles *BGH* and *GHD* also equals two right angles.   I.13 C.N.1

Therefore *a straight line falling on parallel straight lines makes the alternate angles equal to one another, the exterior angle equal to the interior and opposite angle, and the sum of the interior angles on the same side equal to two right angles.*

Q.E.D.

Proposition 30

Straight lines parallel to the same straight line are also parallel to one another.

Let each of the straight lines *AB* and *CD* be parallel to *EF*.

I say that *AB* is also parallel to *CD*.

Let the straight line *GK* fall upon them. Since the straight line *GK* falls on the parallel straight lines *AB* and *EF*, therefore the angle *AGK* equals the angle *GHF*.   I.29

Again, since the straight line *GK* falls on the parallel straight lines *EF* and *CD*, therefore the angle *GHF* equals the angle *GKD*.   I.29

But the angle *AGK* was also proved equal to the angle *GHF*. Therefore the angle *AGK* also equals the angle *GKD*, and they are alternate.   C.N.1

Therefore *AB* is parallel to *CD*.

Therefore *straight lines parallel to the same straight line are also parallel to one another.*

Q.E.D.

Proposition 31

To draw a straight line through a given point parallel to a given straight line.

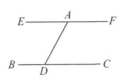

Let *A* be the given point, and *BC* the given straight line. It is required to draw a straight line through the point *A* parallel to the straight line *BC*.

Take a point *D* at random on *BC*. Join *AD*. Construct the angle *DAE* equal to the angle *ADC* on the straight line *DA* and at the point *A* on it. Produce the straight line *AF* in a straight line with *EA*.   Post.1 I.23 Post.2

Since the straight line *AD* falling on the two straight lines *BC* and *EF* makes the alternate angles *EAD* and *ADC* equal to one another, therefore *EAF* is parallel to *BC*.   I.27

Therefore the straight line *EAF* has been drawn through the given point *A* parallel to the given straight line *BC*.

Q.E.F.

Proposition 32

In any triangle, if one of the sides is produced, then the exterior angle equals the sum of the two interior and opposite angles, and the sum of the three interior angles of the triangle equals two right angles.

Let *ABC* be a triangle, and let one side of it *BC* be produced to *D*.

I say that the exterior angle *ACD* equals the sum of the two interior and opposite angles *CAB* and *ABC*, and the sum of the three interior angles of the triangle *ABC*, *BCA*, and *CAB* equals two right angles.

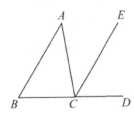

Draw *CE* through the point *C* parallel to the straight line *AB*.   I.31

Since *AB* is parallel to *CE*, and *AC* falls upon them, therefore the alternate angles *BAC* and *ACE* equal one another.   I.29

Again, since *AB* is parallel to *CE*, and the straight line *BD* falls upon them, therefore the exterior angle *ECD* equals the interior and opposite angle *ABC*.   I.29

But the angle *ACE* was also proved equal to the angle *BAC*. Therefore the whole angle *ACD* equals the sum of the two interior and opposite angles *BAC* and *ABC*. Add the angle *ACB* to each. Then the sum of the angles *ACD* and *ACB* equals the sum of the three angles *ABC*, *BCA*, and *CAB*.   C.N.2

But the sum of the angles *ACD* and *ACB* equals two right angles. Therefore the sum of the angles *ABC*, *BCA*, and *CAB* also equals two right angles.   I.13 C.N.1

Therefore *in any triangle, if one of the sides is produced, then the exterior angle equals the sum of the two interior and opposite angles, and the sum of the three interior angles of the triangle equals two right angles.*

Q.E.D.

Exercises on Propositions I.1 – I.32

1. Given: Line CA bisects ∠ BAD
   ∠ CBA = ∠ CDA
   Prove: BC = DC

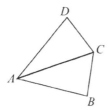

2. Given: AC = BD
   ∠ CAB > ∠ DBA
   Prove: CB > AD

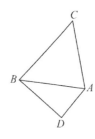

3. In the drawing below ∠ BAC = ∠ CDB. BED and AEC are straight lines and BE = EC.
   Prove: △EAD is isosceles

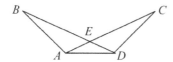

4. Given: DA = DB = DC
   Prove: ∠ ACB is a right angle.

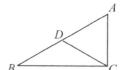

5. Given: In the drawing to the right, ∠ BEC is a right angle and ∠ CAD = ∠ EBC.
   Prove: ∠ ADC is a right angle

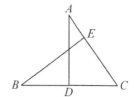

6. Carefully follow the directions:
   a. Draw a semicircle of diameter about four inches and label the endpoints of the diameter A and D.
   b. Without changing the setting of your compass, put the compass point at A and draw an arc that intersects the circumference of your semicircle. Call the point of intersection B.
   c. Again without changing the setting of your compass, put the compass point at B and draw another arc that intersects the circumference of your semicircle. Call this point of intersection C. If you place your compass point at C and draw a third arc in the same manner as you drew the first two, this arc appears to intersect the semicircle at D. Prove that D is, in fact, the point of intersection.

7. Proposition HL

If the hypotenuse and one side of one right triangle is equal to the hypotenuse and one side of another right triangle, respectively, then the remaining side equals the remaining side, the angles are equal respectively, and the triangles are equal.

Let ABC and FDE be two triangles having angles ACB and FED right angles, and let side AC equal side FE and hypotenuse AB equal hypotenuse FD.

I say that $BC = DE$, angle BAC = angle DFE, angle CBA = angle EDF, and triangle ABC = triangle FDE.

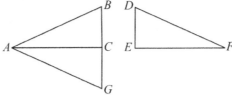

For, let BC be produced to G so that $CG = DE$ and let AG be joined. Then, since angle ACB is right it follows that ACG is also right.	I. 13
Since GC was constructed equal to DE and $AC = EF$ by hypothesis, and since angles ACG and DEF are equal, it follows that $AG = DF$. But $DF = AB$ by hypothesis, hence $AB = AG$.	Post. 4, I. 4, C.N. 1
Now, since triangle BAG is isosceles, angles AGC and ABC are equal.	I.5
And since $AC = AC$ it follows that $BC = CG$.	I.26
But CG was constructed equal to DE. Hence $BC = DE$.	C.N. 1
Thus, the remaining angles are equal respectively and the triangles are equal.	I. 4
	Q.E.D.

Proposition 33

Straight lines which join the ends of equal and parallel straight lines in the same directions are themselves equal and parallel.

Let AB and CD be equal and parallel, and let the straight lines AC and BD join them at their ends in the same directions.

I say that AC and BD are also equal and parallel.

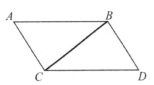

Join BC.	Post.1
Since AB is parallel to CD, and BC falls upon them, therefore the alternate angles ABC and BCD equal one another.	I.29
Since AB equals CD, and BC is common, the two sides AB and BC equal the two sides DC and CB, and the angle ABC equals the angle BCD, therefore the base AC equals the base BD, the triangle ABC equals the triangle DCB, and the remaining angles equals the remaining angles respectively, namely those opposite the equal sides. Therefore the angle ACB equals the angle CBD.	I.4

Since the straight line *BC* falling on the two straight lines *AC* and *BD* makes the alternate angles equal to one another, therefore *AC* is parallel to *BD*.   I.27

And it was also proved equal to it.

Therefore *straight lines which join the ends of equal and parallel straight lines in the same directions are themselves equal and parallel.*

Q.E.D.

Proposition 34

In parallelogrammic areas the opposite sides and angles equal one another, and the diameter bisects the areas.

Let *ACDB* be a parallelogrammic area, and *BC* its diameter. I say that the opposite sides and angles of the parallelogram *ACDB* equal one another, and the diameter *BC* bisects it.

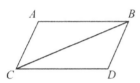

Since *AB* is parallel to *CD*, and the straight line *BC* falls upon them, therefore the alternate angles *ABC* and *BCD* equal one another.   I.29

Again, since *AC* is parallel to *BD*, and *BC* falls upon them, therefore the alternate angles *ACB* and *CBD* equal one another.   I.29

Therefore *ABC* and *DCB* are two triangles having the two angles *ABC* and *BCA* equal to the two angles *DCB* and *CBD* respectively, and one side equal to one side, namely that adjoining the equal angles and common to both of them, *BC*.

Therefore they also have the remaining sides equal to the remaining sides respectively, and the remaining angle to the remaining angle. Therefore the side *AB* equals *CD*, and *AC* equals *BD*, and further the angle *BAC* equals the angle *CDB*.   I.26

Since the angle *ABC* equals the angle *BCD*, and the angle *CBD* equals the angle *ACB*, therefore the whole angle *ABD* equals the whole angle *ACD*.   C.N.2

And the angle *BAC* was also proved equal to the angle *CDB*.

Therefore in parallelogrammic areas the opposite sides and angles equal one another.

I say, next, that the diameter also bisects the areas.

Since *AB* equals *CD*, and *BC* is common, the two sides *AB* and *BC* equal the two sides *DC* and *CB* respectively, and the angle *ABC* equals the angle *BCD*. Therefore the base *AC* also equals *DB*, and the triangle *ABC* equals the triangle *DCB*.   I.4

Therefore the diameter *BC* bisects the parallelogram *ACDB*.

Therefore *in parallelogrammic areas the opposite sides and angles equal one another, and the diameter bisects the areas.*

Q.E.D.

Proposition 35

Parallelograms which are on the same base and in the same parallels equal one another.

Let ABCD and EBCF be parallelograms on the same base BC and in the same parallels AF and BC.

I say that ABCD equals the parallelogram EBCF.

Since ABCD is a parallelogram, therefore AD equals BC.   I.34

For the same reason EF equals BC, so that AD also equals EF.   C.N.1

And DE is common, therefore the whole AE equals the whole DF.   C.N.2

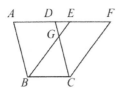

But AB also equals DC. Therefore the two sides EA and AB equal the two sides FD and DC respectively, and the angle FDC equals the angle EAB, the exterior equals the interior. Therefore the base EB equals the base FC, and the triangle EAB equals the triangle FDC.   I.34 / I.29 / I.4

Subtract DGE from each. Then the trapezium ABGD which remains equals the trapezium EGCF which remains.   C.N.3

Add the triangle GBC to each. Then the whole parallelogram ABCD equals the whole parallelogram EBCF.   C.N.2

Therefore parallelograms which are on the same base and in the same parallels equal one another.

Q.E.D.

Proposition 36

Parallelograms which are on equal bases and in the same parallels equal one another.

Let ABCD and EFGH be parallelograms which are on the equal bases BC and FG and in the same parallels AH and BG.

I say that the parallelogram ABCD equals EFGH.

Join BE and CH.   Post.1

Since BC equals FG and FG equals EH, therefore BC equals EH.   I.34 / C.N.1

But they are also parallel, and EB and HC join them. But straight lines joining equal and parallel straight lines in the same directions are equal and parallel, therefore EBCH is a parallelogram.   I.33

And it equals ABCD, for it has the same base BC with it and is in the same parallels BC and AH with it.   I.35

For the same reason also EFGH equals the same EBCH, so that the parallelogram ABCD also equals EFGH.   C.N.1

Therefore parallelograms which are on equal bases and in the same parallels equal one another.

Q.E.D.

Proposition 37

Triangles which are on the same base and in the same parallels equal one another.
Let *ABC* and *DBC* be triangles on the same base *BC* and in the same parallels *AD* and *BC*.
I say that the triangle *ABC* equals the triangle *DBC*.

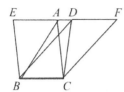

Produce *AD* in both directions to *E* and *F*. Draw *BE* through *B* parallel to *CA*, and draw *CF* through *C* parallel to *BD*.	Post.2 I.31
Then each of the figures *EBCA* and *DBCF* is a parallelogram, and they are equal, for they are on the same base *BC* and in the same parallels *BC* and *EF*.	I.35

Moreover the triangle *ABC* is half of the parallelogram *EBCA*, for the diameter *AB* bisects it. And the triangle *DBC* is half of the parallelogram *DBCF*, for the diameter *DC* bisects it.   I.34

Therefore the triangle *ABC* equals the triangle *DBC*.   C.N

Therefore *triangles which are on the same base and in the same parallels equal one another.*

Q.E.D.

Proposition 38

Triangles which are on equal bases and in the same parallels equal one another.
Let *ABC* and *DEF* be triangles on equal bases *BC* and *EF* and in the same parallels *BF* and *AD*.
I say that the triangle *ABC* equals the triangle *DEF*.

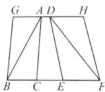

Produce *AD* in both directions to *G* and *H*. Draw *BG* through *B* parallel to *CA*, and draw *FH* through *F* parallel to *DE*.	Post.2 I.31
Then each of the figures *GBCA* and *DEFH* is a parallelogram, and *GBCA* equals *DEFH*, for they are on equal bases *BC* and *EF* and in the same parallels *BF* and *GH*.	I.36

Moreover the triangle *ABC* is half of the parallelogram *GBCA*, for the diameter *AB* bisects it. And the triangle *FED* is half of the parallelogram *DEFH*, for the diameter *DF* bisects it.   I.34

Therefore the triangle *ABC* equals the triangle *DEF*.   C.N.

Therefore triangles which are on equal bases and in the same parallels equal one another.

Q.E.D.

Proposition 39

Equal triangles which are on the same base and on the same side are also in the same parallels.

Let *ABC* and *DBC* be equal triangles which are on the same base *BC* and on the same side of it.
Join *AD*.
I say that *AD* is parallel to *BC*.

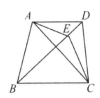

If not, draw *AE* through the point *A* parallel to the straight line *BC*, and join *EC*.	I.31 Post.1
Therefore the triangle *ABC* equals the triangle *EBC*, for it is on the same base *BC* with it and in the same parallels.	I.37
But *ABC* equals *DBC*, therefore *DBC* also equals *EBC*, the greater equals the less, which is impossible.	C.N.1

Therefore *AE* is not parallel to *BC*.
Similarly we can prove that neither is any other straight line except *AD*, therefore *AD* is parallel to *BC*.

Therefore *equal triangles which are on the same base and on the same side are also in the same parallels.*

Q.E.D.

Proposition 40

Equal triangles which are on equal bases and on the same side are also in the same parallels.

Let *ABC* and *CDE* be equal triangles on equal bases *BC* and *CE* and on the same side.
I say that they are also in the same parallels.

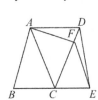

Join *AD*. I say that *AD* is parallel to *BE*.	Post.1
If not, draw *AF* through *A* parallel to *BE*, and join *FE*.	I.31 Post.1
Therefore the triangle *ABC* equals the triangle *FCE*, for they are on equal bases *BC* and *CE* and in the same parallels *BE* and *AF*.	I.38

But the triangle *ABC* equals the triangle *DCE*, therefore the triangle *DCE* also equals the triangle *FCE*, the greater equals the less, which is impossible. Therefore *AF* is not parallel to *BE*.   C.N.1

Similarly we can prove that neither is any other straight line except *AD*, therefore *AD* is parallel to *BE*.

Therefore *equal triangles which are on equal bases and on the same side are also in the same parallels.*

Q.E.D.

Proposition 41

If a parallelogram has the same base with a triangle and is in the same parallels, then the parallelogram is double the triangle.

Let the parallelogram ABCD have the same base BC with the triangle EBC, and let it be in the same parallels BC and AE.

I say that the parallelogram ABCD is double the triangle BEC.

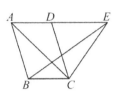

Join AC.	Post.1
Then the triangle ABC equals the triangle EBC, for it is on the same base BC with it and in the same parallels BC and AE.	I.37
But the parallelogram ABCD is double the triangle ABC, for the diameter AC bisects it, so that the parallelogram ABCD is also double the triangle EBC.	I.34

Therefore *if a parallelogram has the same base with a triangle and is in the same parallels, then the parallelogram is double the triangle.*

Q.E.D.

Proposition 42

To construct a parallelogram equal to a given triangle in a given rectilinear angle.

Let ABC be the given triangle, and D the given rectilinear angle.

It is required to construct a parallelogram equal to the triangle ABC in the rectilinear angle D.

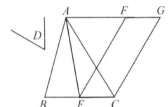

Bisect BC at E, and join AE. Construct the angle CEF on the straight line EC at the point E on it equal to the angle D. Draw AG through A parallel to EC, and draw CG through C parallel to EF.	I.10 Post. 1 I.23 I.31
Then FECG is a parallelogram.	

Since BE equals EC, therefore the triangle ABE also equals the triangle AEC, for they are on equal bases BE and EC and in the same parallels BC and AG. Therefore the triangle ABC is double the triangle AEC.	I.38
But the parallelogram FECG is also double the triangle AEC, for it has the same base with it and is in the same parallels with it, therefore the parallelogram FECG equals the triangle ABC.	I.41 C.N. 1

And it has the angle CEF equal to the given angle D.

Therefore the parallelogram FECG has been constructed equal to the given triangle ABC, in the angle CEF which equals D.

Q.E.F.

Proposition 43

In any parallelogram the complements of the parallelograms about the diameter equal one another.

Let *ABCD* be a parallelogram, and *AC* its diameter, and about *AC* let *EH* and *FG* be parallelograms, and *BK* and *KD* the so-called complements.
I say that the complement *BK* equals the complement *KD*.

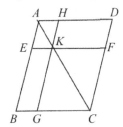

Since *ABCD* is a parallelogram, and *AC* its diameter, therefore the triangle *ABC* equals the triangle *ACD*.   I.34

Again, since *EH* is a parallelogram, and *AK* is its diameter, therefore the triangle *AEK* equals the triangle *AHK*. For the same reason the triangle *KFC* also equals *KGC*.   I.34

Now, since the triangle *AEK* equals the triangle *AHK*, and *KFC* equals *KGC*, therefore the triangle *AEK* together with *KGC* equals the triangle *AHK* together with *KFC*.   C.N. 2

And the whole triangle *ABC* also equals the whole *ADC*, therefore the remaining complement *BK* equals the remaining complement *KD*.   C.N. 3

Therefore *in any parallelogram the complements of the parallelograms about the diameter equal one another.*

Q.E.D.

Proposition 44

To a given straight line, in a given rectilinear angle, to apply a parallelogram equal to a given triangle.

Let *AB* be the given straight line, *D* the given rectilinear angle, and *C* the given triangle.
It is required to apply a parallelogram equal to the given triangle *C* to the given straight line *AB* in an angle equal to *D*.

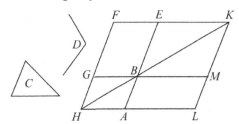

Construct the parallelogram *BEFG* equal to the triangle *C* in the angle *EBG* which equals *D*, and let it be placed so that *BE* is in a straight line with *AB*.   I.42

Draw *FG* through to *H*, and draw *AH* through *A* parallel to either *BG* or *EF*.
Join *HB*.   I.31

Since the straight line *HF* falls upon the parallels *AH* and *EF*, therefore the sum of the angles *AHF* and *HFE* equals two right angles. Therefore the sum of the angles *BHG* and *GFE* is less than two right angles. And straight lines produced indefinitely from angles less than two right angles meet, therefore *HB* and *FE*, when produced, will meet.   I.29 Post.5

Let them be produced and meet at *K*. Draw *KL* through the point *K* parallel to either *EA* or *FH*. Produce *HA* and *GB* to the points *L* and *M*.	I.31
Then *HLKF* is a parallelogram, *HK* is its diameter, and *AG* and *ME* are parallelograms, and *LB* and *BF* are the so-called complements about *HK*. Therefore *LB* equals *BF*.	I.43
But *BF* equals the triangle *C*, therefore *LB* also equals *C*.	C.N.1
Since the angle *GBE* equals the angle *ABM*, while the angle *GBE* equals *D*, therefore the angle *ABM* also equals the angle *D*.	I.15 C.N.1
Therefore the parallelogram *LB* equal to the given triangle *C* has been applied to the given straight line *AB*, in the angle *ABM* which equals *D*.	
	Q.E.F.

Proposition 45

To construct a parallelogram equal to a given rectilinear figure in a given rectilinear angle.
Let *ABCD* be the given rectilinear figure and *E* the given rectilinear angle.
It is required to construct a parallelogram equal to the rectilinear figure *ABCD* in the given angle *E*.

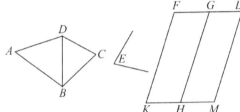

Join *DB*. Construct the parallelogram *FH* equal to the triangle *ABD* in the angle *HKF* which equals *E*. Apply the parallelogram *GM* equal to the triangle *DBC* to the straight line *GH* in the angle *GHM* which equals *E*.	I.42 I.44
Since the angle *E* equals each of the angles *HKF* and *GHM*, therefore the angle *HKF* also equals the angle *GHM*.	C.N.1
Add the angle *KHG* to each. Therefore the sum of the angles *FKH* and *KHG* equals the sum of the angles *KHG* and *GHM*.	C.N.2
But the sum of the angles *FKH* and *KHG* equals two right angles, therefore the sum of the angles *KHG* and *GHM* also equals two right angles.	I.29 C.N.1
Thus, with a straight line *GH*, and at the point *H* on it, two straight lines *KH* and *HM* not lying on the same side make the adjacent angles together equal to two right angles, therefore *KH* is in a straight line with *HM*.	I.14
Since the straight line *HG* falls upon the parallels *KM* and *FG*, therefore the alternate angles *MHG* and *HGF* equal one another.	I.29
Add the angle *HGL* to each. Then the sum of the angles *MHG* and *HGL* equals the sum of the angles *HGF* and *HGL*.	C.N.2

But the sum of the angles *MHG* and *HGL* equals two right angles, therefore the sum of the angles *HGF* and *HGL* also equals two right angles. Therefore *FG* is in a straight line with *GL*.

I.29
C.N.1
I.14

Since *FK* is equal and parallel to *HG*, and *HG* equal and parallel to *ML* also, therefore *KF* is also equal and parallel to *ML*, and the straight lines *KM* and *FL* join them at their ends. Therefore *KM* and *FL* are also equal and parallel. Therefore *KFLM* is a parallelogram.

I.34
I.30
C.N.1
I.33

Since the triangle *ABD* equals the parallelogram *FH*, and *DBC* equals *GM*, therefore the whole rectilinear figure *ABCD* equals the whole parallelogram *KFLM*.

C.N.2

Therefore the parallelogram *KFLM* has been constructed equal to the given rectilinear figure *ABCD* in the angle *FKM* which equals the given angle *E*.

Q.E.F.

Proposition 46

To describe a square on a given straight line.

Let *AB* be the given straight line.
It is required to describe a square on the straight line *AB*.

Draw *AC* at right angles to the straight line *AB* from the point *A* on it. Make *AD* equal to *AB*. Draw *DE* through the point *D* parallel to *AB*, and draw *BE* through the point *B* parallel to *AD*.

I.11
I.3
I.31

Then *ADEB* is a parallelogram. Therefore *AB* equals *DE*, and *AD* equals *BE*.

I.34

But *AB* equals *AD*, therefore the four straight lines *BA*, *AD*, *DE*, and *EB* equal one another. Therefore the parallelogram *ADEB* is equilateral.

I say next that it is also right-angled.

Since the straight line *AD* falls upon the parallels *AB* and *DE*, therefore the sum of the angles *BAD* and *ADE* equals two right angles.

I.29

But the angle *BAD* is right, therefore the angle *ADE* is also right.

And in parallelogrammic areas the opposite sides and angles equal one another, therefore each of the opposite angles *ABE* and *BED* is also right. Therefore *ADEB* is right-angled.

I.34

And it was also proved equilateral.

Therefore it is a square, and it is described on the straight line *AB*.

I.Def.22

Q.E.F.

Proposition 47

In right-angled triangles the square on the side opposite the right angle equals the sum of the squares on the sides containing the right angle.

Let ABC be a right-angled triangle having the angle BAC right.

I say that the square on BC equals the sum of the squares on BA and AC.

Describe the square BDEC on BC, and the squares GB and HC on BA and AC. Draw AL through A parallel to either BD or CE, and join AD and FC.

I.46
I.31

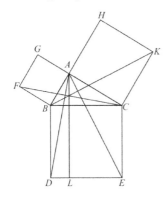

Since each of the angles BAC and BAG is right, it follows that with a straight line BA, and at the point A on it, the two straight lines AC and AG not lying on the same side make the adjacent angles equal to two right angles, therefore CA is in a straight line with AG.

I.Def.22
I.14

For the same reason BA is also in a straight line with AH.

Since the angle DBC equals the angle FBA, for each is right, add the angle ABC to each, therefore the whole angle DBA equals the whole angle FBC.

I.Def.22
Post.4
C.N.2

Since DB equals BC, and FB equals BA, the two sides AB and BD equal the two sides FB and BC respectively, and the angle ABD equals the angle FBC, therefore the base AD equals the base FC, and the triangle ABD equals the triangle FBC.

I.Def.22
I.4

Now the parallelogram BL is double the triangle ABD, for they have the same base BD and are in the same parallels BD and AL. And the square GB is double the triangle FBC, for they again have the same base FB and are in the same parallels FB and GC.

I.41

Therefore the parallelogram BL also equals the square GB.

Similarly, if AE and BK are joined, the parallelogram CL can also be proved equal to the square HC. Therefore the whole square BDEC equals the sum of the two squares GB and HC.

C.N.2

And the square BDEC is described on BC, and the squares GB and HC on BA and AC.

Therefore the square on BC equals the sum of the squares on BA and AC.

Therefore in right-angled triangles the square on the side opposite the right angle equals the sum of the squares on the sides containing the right angle.

Q.E.D.

Proposition 48

If in a triangle the square on one of the sides equals the sum of the squares on the remaining two sides of the triangle, then the angle contained by the remaining two sides of the triangle is right.

In the triangle ABC let the square on one side BC equal the sum of the squares on the sides BA and AC

I say that the angle BAC is right.

Draw AD from the point A at right angles to the straight line AC. Make AD equal to BA, and join DC.	I.11 I.3
Since DA equals AB, therefore the square on DA also equals the square on AB.	
Add the square on AC to each. Then the sum of the squares on DA and AC equals the sum of the squares on BA and AC.	C.N.2

But the square on DC equals the sum of the squares on DA and AC, for the angle DAC is right, and the square on BC equals the sum of the squares on BA and AC, for this is the hypothesis, therefore the square on DC equals the square on BC, so that the side DC also equals BC.

Since DA equals AB, and AC is common, the two sides DA and AC equal the two sides BA and AC, and the base DC equals the base BC, therefore the angle DAC equals the angle BAC. But the angle DAC is right, therefore the angle BAC is also right.

I.47
C.N.1

I.8

Therefore *if in a triangle the square on one of the sides equals the sum of the squares on the remaining two sides of the triangle, then the angle contained by the remaining two sides of the triangle is right.*

Q.E.D.

Exercises on Book I

1. Given: BCDE is a straight line
   Prove: ∠EFD < ∠ACB

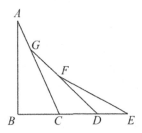

2. Given: AB = AC
   BCD is a straight line
   Prove: AD > AB

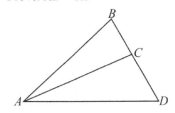

3. Given: △ABC is isosceles
∠AED = ∠ABC
Prove: DE ∥ BC

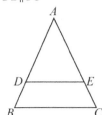

4. Given quadrilateral ABCD with exterior angles FBC and GDC.

Prove that the sum of the angles FBC and GDC is equal to the sum of the angles BAD and BCD.

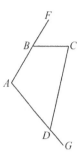

5. Prove that the diagonals of a parallelogram bisect one another.

6. Given: △ABC with point E bisecting line AC
Follow the directions:
   a. Draw a line through E parallel to BC. Let D be the point of intersection of this line with AB
   b. Draw a line through C parallel to BA above BC
   c. Produce DE until it intersects this line at F
   d. Prove △ADE and △CFE are equal, have equal sides respectively, and equal angles respectively
   e. Prove point D bisects line AB

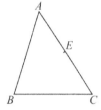

7. Given: AB = AC
BE = ED
BE bisects ∠ABC
Prove: △AED = △BEC

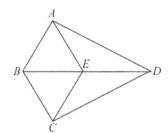

8. Given: AB = AC
AD ⊥ BC
Prove: △BDE = △CDE

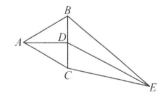

9. Given: △ABD ≅ △ACD
   ∠BAC = ∠ACD
   Prove: ABCD is a parallelogram

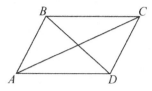

10. Given: AE = EB
    ∠CAE = ∠DBE
    Prove: CE < ED

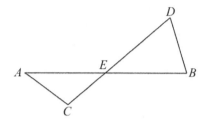

11. Given: BC ⊥ AD
    AB = CD
    ∠BAD and ∠CDA are both acute
    Prove: ∠BAD = ∠CDA

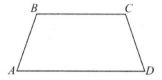

12. Given: ABCD is a parallelogram.
    ∠DFC and ∠AEB are right angles
    Prove: The sum of the squares on AE and EB is equal to the sum of the squares on DF and FC

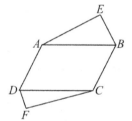

13. In the figures to the right, the largest are squares whose sides are equal in length. As you can see, the sides of these large squares are each divided into two sections, one longer than the other. The longer sections of all eight sides are equal. (Which means, of course, that the smaller sections are all equal as well.) Answer the following questions, and then use your answers to develop an alternate proof of the Pythagorean Theorem.

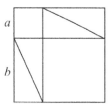

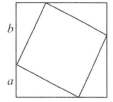

 a. What can you say about the areas of the large squares, compared to one another?
 b. What can you say about the area of the triangles in the figure on the top compared to the area of the triangles in the figure on the bottom?
 c. Prove that the figure in the middle of the square on the bottom is a square.
 d. Inside the top large square are two figures that appear to be squares. Prove that they are, in fact, squares.
 e. Using your answers to parts a. through d. above, prove the Pythagorean Theorem.

Book III

Definitions

1. *Equal circles* are those whose diameters are equal, or whose radii are equal.

2. A straight line is said to *touch* a circle which, meeting the circle and being produced, does not cut the circle.

3. Circles are said to *touch* one another which meet one another but do not cut one another.

4. Straight lines in a circle are said to be *equally distant* from the center when the perpendiculars drawn to them from the center are equal.

5. And that straight line is said to be at a *greater distance* on which the greater perpendicular falls.

6. A *segment* of a circle is the figure contained by a straight line and a circumference of a circle.

7. An *angle of a segment* is that contained by a straight line and a circumference of a circle.

8. An *angle in a segment* is the angle which, when a point is taken on the circumference of the segment and straight lines are joined from it to the ends of the straight line which is the base of the segment, is contained by the straight lines so joined.

9. And, when the straight lines containing the angle cut off a circumference, the angle is said to *stand upon* that circumference.

10. A *sector* of a circle is the figure which, when an angle is constructed at the center of the circle, is contained by the straight lines containing the angle and the circumference cut off by them.

11. *Similar segments* of circles are those which admit equal angles, or in which the angles equal one another.

Book III Propositions

Proposition 1

To find the center of a given circle.
Let *ABC* be the given circle.
It is required to find the center of the circle *ABC*.

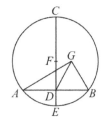

Draw a straight line *AB* through it at random, and bisect it at the point *D*. Draw *DC* from *D* at right angles to *AB*, and draw it through to *E*. Bisect *CE* at *F*.

I.10
I.11
I.10

I say that *F* is the center of the circle *ABC*.

For suppose it is not, but, if possible, let *G* be the center. Join *GA*, *GD*, and *GB*.

Then, since *AD* equals *DB*, and *DG* is common, the two sides *AD* and *DG* equal the two sides *BD* and *DG* respectively. And the base *GA* equals the base *GB*, for they are radii, therefore the angle *ADG* equals the angle *GDB*.

I.Def.15
I.8

But, when a straight line standing on a straight line makes the adjacent angles equal to one another, each of the equal angles is right, therefore the angle *GDB* is right.

I.Def.10

But the angle *FDB* is also right, therefore the angle *FDB* equals the angle *GDB*, the greater equals the less, which is impossible. Therefore *G* is not the center of the circle *ABC*.

Similarly we can prove that neither is any other point except *F*.

Therefore the point *F* is the center of the circle *ABC*.

Q.E.F.

Corollary

From this it is manifest that *if in a circle a straight line cuts a straight line into two equal parts and at right angles, then the center of the circle lies on the cutting straight line.*

Proposition 2

If two points are taken at random on the circumference of a circle, then the straight line joining the points falls within the circle.

Let ABC be a circle, and let two points A and B be taken at random on its circumference.
I say that the straight line joined from A to B falls within the circle.

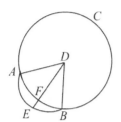

For suppose it does not, but, if possible, let it fall outside, as AEB. Take the center D of the circle ABC. Join DA and DB, and draw DFE through.   III.1

Then, since DA equals DB, the angle DAE also equals the angle DBE.   I.Def.15   I.5

And, since one side AEB of the triangle DAE is produced, the angle DEB is greater than the angle DAE.   I.16

And the angle DAE equals the angle DBE, therefore the angle DEB is greater than the angle DBE. And the side opposite the greater angle is greater, therefore DB is greater than DE. But DB equals DF, therefore DF is greater than DE, the less greater than the greater, which is impossible.   I.19   I.Def.15

Therefore the straight line joined from A to B does not fall outside the circle. Similarly we can prove that neither does it fall on the circumference itself, therefore it falls within.

Therefore *if two points are taken at random on the circumference of a circle, then the straight line joining the points falls within the circle.*

Q.E.D.

Proposition 3

If a straight line passing through the center of a circle bisects a straight line not passing through the center, then it also cuts it at right angles; and if it cuts it at right angles, then it also bisects it.

Let a straight line CD passing through the center of a circle ABC bisect a straight line AB not passing through the center at the point F.
I say that it also cuts it at right angles.

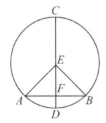

Take the center E of the circle ABC, and join EA and EB.   III.1

Then, since AF equals FB, and FE is common, two sides equal two sides, and the base EA equals the base EB, therefore the angle AFE equals the angle BFE.   I.Def.15   I.8

But, when a straight line standing on another straight line makes the adjacent angles equal to one another, each of the equal angles is right, therefore each of the angles AFE and BFE is right.   I.Def.10

Therefore *CD*, which passes through the center and bisects *AB* which does not pass through the center, also cuts it at right angles.
Next, let *CD* cut *AB* at right angles.
I say that it also bisects it, that is, that *AF* equals *FB*.

For, with the same construction, since *EA* equals *EB*, the angle *EAF* also equals the angle *EBΓ*.   I.5

But the right angle *AFE* equals the right angle *BFE*, therefore *EAF* and *EBF* are two triangles having two angles equal to two angles and one side equal to one side, namely *EF*, which is common to them, and opposite one of the equal angles. Therefore they also have the remaining sides equal to the remaining sides   I.26

Therefore *AF* equals *FB*.

Therefore *if a straight line passing through the center of a circle bisects a straight line not passing through the center, then it also cuts it at right angles; and if it cuts it at right angles, then it also bisects it.*

Q.E.D.

Proposition 4

If in a circle two straight lines which do not pass through the center cut one another, then they do not bisect one another.

Let *ABCD* be a circle, and in it let the two straight lines *AC* and *BD*, which do not pass through the center, cut one another at *E*.
I say that they do not bisect one another.

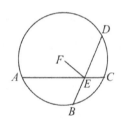

For, if so, let them bisect one another, so that *AE* equals *EC*, and *BE* equals *ED*. Take the center *F* of the circle *ABCD*. Join *FE*.   III.1

Then, since a straight line *FE* passing through the center bisects a straight line *AC* not passing through the center, it also cuts it at right angles, therefore the angle *FEA* is right.   III.3

Again, since a straight line *FE* bisects a straight line *BD*, it also cuts it at right angles. Therefore the angle *FEB* is right.   III.3

But the angle *FEA* was also proved right, therefore the angle *FEA* equals the angle *FEB*, the less equals the greater, which is impossible.
Therefore *AC* and *BD* do not bisect one another.

Therefore *if in a circle two straight lines which do not pass through the center cut one another, then they do not bisect one another.*

Q.E.D.

Proposition 5

If two circles cut one another, then they do not have the same center.

Let the circles ABC and CDG cut one another at the points B and C.
I say that they do not have the same center.

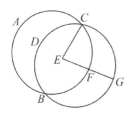

For, if possible, let it be E. Join EC, and draw EFG through at random.

Then, since the point E is the center of the circle ABC, EC equals EF. Again, since the point E is the center of the circle CDG, EC equals EG. I.Def.15

But EC was proved equal to EF also, therefore EF also equals EG, the less equals the greater which is impossible. I.Def.15

Therefore the point E is not the center of the circles ABC and CDG.

Therefore *if two circles cut one another, then they do not have the same center.*

Q.E.D.

Proposition 6

If two circles touch one another, then they do not have the same center.

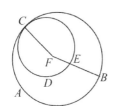

Let the two circles ABC and CDE touch one another at the point C.
I say that they do not have the same center.
For, if possible, let it be F. Join FC, and draw FEB through at random.

Then, since the point F is the center of the circle ABC, FC equals FB. Again, since the point F is the center of the circle CDE, FC equals FE. I.Def.15

But FC was proved equal to FB, therefore FE also equals FB, the less equals the greater, which is impossible.

Therefore F is not the center of the circles ABC and CDE.

Therefore *if two circles touch one another, then they do not have the same center.*

Q.E.D.

Proposition 7

If on the diameter of a circle a point is taken which is not the center of the circle, and from the point straight lines fall upon the circle, then that is greatest on which passes through the center, the remainder of the same diameter is least, and of the rest the nearer to the straight line through the center is always greater than the more remote; and only two equal straight lines fall from the point on the circle, one on each side of the least straight line.

Let *ABCD* be a circle, and let *AD* be a diameter of it. Let *F* be a point *F* on *AD* which is not the center of the circle. Let *E* be the center of the circle. Let straight lines *FB*, *FC*, and *FG* fall upon the circle *ABCD* from *F*.

I say that *FA* is greatest, *FD* is least, and of the rest *FB* is greater than *FC*, and *FC* greater than *FG*.

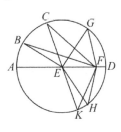

Join *BE*, *CE*, and *GE*.

Then, since in any triangle the sum of any two sides is greater than the remaining one, the sum of *EB* and *EF* is greater than *BF*.   I.20

But *AE* equals *BE*, therefore *AF* is greater than *BF*.

Again, since *BE* equals *CE*, and *FE* is common, the two sides *BE* and *EF* equal the two sides *CE* and *EF*. But the angle *BEF* is also greater than the angle *CEF*, therefore the base *BF* is greater than the base *CF*.   I.24

For the same reason *CF* is also greater than *GF*.

Again, since the sum of *GF* and *FE* is greater than *EG*, and *EG* equals *ED*, the sum of *GF* and *FE* is greater than *ED*.   I.20

Subtract *EF* from each. Therefore the remainder *GF* is greater than the remainder *FD*. Therefore *FA* is greatest, *FD* is least, *FB* is greater than *FC*, and *FC* greater than *FG*.

I say also that from the point *F* only two equal straight lines fall on the circle *ABCD*, one on each side of the least *FD*.

Construct the angle *FEH* equal to the angle *GEF* on the straight line *EF* and at the point *E* on it. Join *FH*.   I.23

Then, since *GE* equals *EH*, and *EF* is common, the two sides *GE* and *EF* equal the two sides *HE* and *EF*, and the angle *GEF* equals the angle *HEF*, therefore the base *FG* equals the base *FH*.   I.4

I say again that another straight line equal to *FG* does not fall on the circle from the point *F*.

For, if possible, let *FK* so fall.

Then, since *FK* equals *FG*, and *FH* equals *FG*, *FK* also equals *FH*, the nearer to the straight line through the center being thus equal to the more remote, which is impossible.

Therefore another straight line equal to *GF* does not fall from the point *F* upon the circle. Therefore only one straight line so falls.

Therefore *if on the diameter of a circle a point is taken which is not the center of the circle, and from the point straight lines fall upon the circle, then that is greatest on which passes through the center, the remainder of the same diameter is least, and of the rest the nearer to the straight line through the center is always greater than the more remote; and only two equal straight lines fall from the point on the circle, one on each side of the least straight line.*

                                                                                                                        Q.E.D.

Proposition 8

If a point is taken outside a circle and from the point straight lines are drawn through to the circle, one of which is through the center and the others are drawn at random, then, of the straight lines which fall on the concave circumference, that through the center is greatest, while of the rest the nearer to that through the center is always greater than the more remote, but, of the straight lines falling on the convex circumference, that between the point and the diameter is least, while of the rest the nearer to the least is always less than the more remote; and only two equal straight lines fall on the circle from the point, one on each side of the least.

Let ABC be a circle, and let a point D be taken outside ABC. Let straight lines DA, DE, DF, and DC be drawn through from D, and let DA be drawn through the center. I say that, of the straight lines falling on the concave circumference AEFC, the straight line DA through the center is greatest, while DE is greater than DF, and DF greater than DC. But, of the straight lines falling on the convex circumference HLKG, the straight line DG between the point and the diameter AG is least, and the nearer to the least DG is always less than the more remote, namely DK is less than DL, and DL is less than DH.

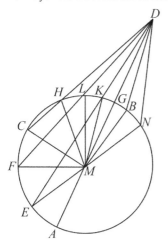

Take the center M of the circle ABC. Join ME, MF, MC, MK, ML, and MH. — III.1

Then, since AM equals EM, add MD to each, therefore AD equals the sum of EM and MD.

But the sum of EM and MD is greater than ED, therefore AD is also greater than ED. — I.20

Again, since ME equals MF, and MD is common, therefore EM and MD equal FM and MD, and the angle EMD is greater than the angle FMD, therefore the base ED is greater than the base FD. — I.24

Similarly we can prove that FD is greater than CD. Therefore DA is greatest, while DE is greater than DF, and DF is greater than DC.

Next, since the sum of MK and KD is greater than MD, and MG equals MK, therefore the remainder KD is greater than the remainder GD, so that GD is less than KD. — I.20

And, since on MD, one of the sides of the triangle MLD, two straight lines MK and KD are constructed meeting within the triangle, therefore the sum of MK and KD is less than the sum of ML and LD. And MK equals ML, therefore the remainder DK is less than the remainder DL. — I.21

Similarly we can prove that DL is also less than DH. Therefore DG is least, while DK is less than DL, and DL is less than DH.

I say also that only two equal straight lines will fall from the point D on the circle, one on each side of the least DG.

Construct the angle *DMB* equal to the angle *KMD* on the straight line *MD* and at the point *M* on it. Join *DB*.   I.23

Then, since *MK* equals *MB*, and *MD* is common, the two sides *KM* and *MD* equal the two sides *BM* and *MD* respectively, and the angle *KMD* equals the angle *BMD*, therefore the base *DK* equals the base *DB*.   I.4

I say that no other straight line equal to the straight line *DK* falls on the circle from the point *D*.

For, if possible, let a straight line so fall, and let it be *DN*. Then, since *DK* equals *DN*, and *DK* equals *DB*, *DB* also equals *DN*, that is, the nearer to the least *DG* equal to the more remote, which was proved impossible.

Therefore no more than two equal straight lines fall on the circle *ABC* from the point *D*, one on each side of *DG* the least.

Therefore *if a point is taken outside a circle and from the point straight lines are drawn through to the circle, one of which is through the center and the others are drawn at random, then, of the straight lines which fall on the concave circumference, that through the center is greatest, while of the rest the nearer to that through the center is always greater than the more remote, but, of the straight lines falling on the convex circumference, that between the point and the diameter is least, while of the rest the nearer to the least is always less than the more remote; and only two equal straight lines fall on the circle from the point, one on each side of the least.*

Q.E.D.

Proposition 9

If a point is taken within a circle, and more than two equal straight lines fall from the point on the circle, then the point taken is the center of the circle.

Let *D* a point within a circle *ABC*, and from *D* let more than two equal straight lines, namely *DA* and *DB* and *DC*, fall on the circle *ABC*.
I say that the point *D* is the center of the circle *ABC*.

Join *AB* and *BC*, and bisect them at the points *E* and *F*. Join *ED* and *FD*, and draw them through to the points *G*, *K*, *H*, and *L*.

Then, since *AE* equals *EB*, and *ED* is common, the two sides *AE* and *ED* equal the two sides *BE* and *ED*, and the base *DA* equals the base *DB*, therefore the angle *AED* equals the angle *BED*.   I.8

Therefore the angles *AED* and *BED* are each right. Therefore *GK* cuts *AB* into two equal parts and at right angles.   I. Def. 10

And since, if in a circle a straight line cuts a straight line into two equal parts and at right angles, the center of the circle is on the cutting straight line, therefore the center of the circle is on *GK*.   III.1, Cor.

For the same reason the center of the circle *ABC* is also on *HL*.

And the straight lines GK and HL have no other point common but the point D, therefore the point D is the center of the circle ABC.

Therefore *if a point is taken within a circle, and more than two equal straight lines fall from the point on the circle, then the point taken is the center of the circle.*

Q.E.D.

Proposition 10

A circle does not cut a circle at more than two points.

For, if possible, let the circle ABC cut the circle DEF at more points than two, namely B, G, F, and H.

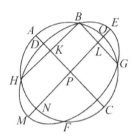

Join BH and BG, and bisect them at the points K and L. Draw KC and LM from K and L at right angles to BH and BG, and carry them through to the points A and E.

Then, since in the circle ABC a straight line AC cuts a straight line BH into two equal parts and at right angles, the center of the circle ABC lies on AC. Again, since in the same circle ABC a straight line NO cuts a straight line BG into two equal parts and at right angles, the center of the circle ABC lies on NO.

III.1,Cor.

But it was also proved to lie on AC, and the straight lines AC and NO meet at no point except at P, therefore the point P is the center of the circle ABC.

Similarly we can prove that P is also the center of the circle DEF, therefore the two circles ABC and DEF which cut one another have the same center P, which is impossible.

III.5

Therefore *a circle does not cut a circle at more than two points.*

Q.E.D.

Proposition 11

If two circles touch one another internally, and their centers are taken, then the straight line joining their centers, being produced, falls on the point of contact of the circles.

Let the two circles ABC and ADE touch one another internally at the point A, and let the centers F and G of the circles ABC and ADE be taken.

I say that the straight line joined from G to F and produced falls on A.

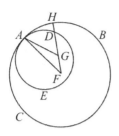

For suppose it does not, but, if possible, let it fall as *FGH*.
Join *AF* and *AG*.

Then, since the sum of *AG* and *GF* is greater than *FA*, that is, than *FH*, subtract *FG* from each, therefore the remainder *AG* is greater than the remainder *GH*.

I.20

But *AG* equals *GD*, therefore *GD* is also greater than *GH*, the less greater than the greater, which is impossible.

Therefore the straight line joined from *F* to *G* does not fall outside. Therefore it falls on *A*, the point of contact.

Therefore *if two circles touch one another internally, and their centers are taken, then the straight line joining their centers, being produced, falls on the point of contact of the circles.*

Q.E.D.

Proposition 12

If two circles touch one another externally, then the straight line joining their centers passes through the point of contact.

Let the two circles *ABC* and *ADE* touch one another externally at the point *A*. Take the center *F* of *ABC*, and the center *G* of *ADE*.

III.1

I say that the straight line joined from *F* to *G* passes through the point of contact at *A*.

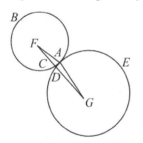

For suppose it does not, but, if possible, let it pass as *FCDG*. Join *AF* and *AG*.
Then, since the point *F* is the center of the circle *ABC*, *FA* equals *FC*.
Again, since the point *G* is the center of the circle *ADE*, *GA* equals *GD*.
But *FA* was also proved equal to *FC*, therefore *FA* and *AG* equal *FC* and *GD*, so that the whole *FG* is greater than *FA* and *AG*, but it is also less, which is impossible.

I.20

Therefore the straight line joined from *F* to *G* does not fail to pass through the point of contact at *A*, therefore it passes through it.

Therefore *if two circles touch one another externally, then the straight line joining their centers passes through the point of contact.*

Q.E.D.

Proposition 13

A circle does not touch another circle at more than one point whether it touches it internally or externally.

For, if possible, let the circle ABDC touch the circle EBFD, first internally, at more points than one, namely D and B.

Take the center G of the circle ABDC and the center H of EBFD.

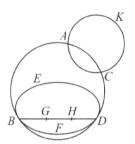

Therefore the straight line joined from G to H falls on B and D. Let it so fall, as BGHD.   III.11

Then, since the point G is the center of the circle ABCD and BG equals GD, therefore BG is greater than HD. Therefore BH is much greater than HD.

Again, since the point H is the center of the circle EBFD, BH equals HD, but it was also proved much greater than it, which is impossible.

Therefore a circle does not touch a circle internally at more points than one.

I say further that neither does it so touch it externally.

For, if possible, let the circle ACK touch the circle ABDC at more points than one, namely A and C and let AC be joined. Then, since on the circumference of each of the circles ABDC and ACK two points A and C have been taken at random, the straight line joining the points falls within each circle;   III.2

but it fell within the circle ABCD and outside ACK, which is absurd.   III. Def.3

Therefore, a circle does not touch a circle externally at more points than one. And it was proved that neither does it so touch it internally.

Therefore, *a circle does not touch another circle at more than one point whether it touches it internally or externally.*

                                                                                                          Q.E.D.

Proposition 14

Equal straight lines in a circle are equally distant from the center, and those which are equally distant from the center equal one another.

Let AB and CD be equal straight lines in a circle ABDC.
I say that AB and CD are equally distant from the center.

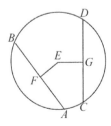

Take the center E of the circle ABDC. Draw EF and EG from E perpendicular to AB and CD, and join AE and EC.   III.1   I.12

Then, since a straight line EF passing through the center cuts a straight line AB not passing through the center at right angles, it also bisects it. Therefore AF equals FB. Therefore AB is double AF.   III.3

For the same reason *CD* is also double *CG*. But *AB* equals *CD*, therefore *AF* also equals *CG*.

Also, since *AE* equals *EC*, the square on *AE* also equals the square on *EC*. But the sum of the squares on *AF* and *EF* equals the square on *AE*, for the angle at *F* is right, and the sum of the squares on *EG* and *GC* equals the square on *EC*, for the angle at *G* is right. Therefore the sum of the squares on *AF* and *FE* equals the sum of the squares on *CG* and *GE*, of which the square on *AF* equals the square on *CG*, for *AF* equals *CG*. Therefore the remaining square on *FE* equals the square on *EG*. Therefore *EF* equals *EG*.

I.47

But straight lines in a circle are said to be equally distant from the center when the perpendiculars drawn to them from the center are equal. Therefore *AB* and *CD* are equally distant from the center.

III.Def.4

Next, let the straight lines *AB* and *CD* be equally distant from the center, that is, let *EF* equal *EG*.
I say that *AB* also equals *CD*.

For, with the same construction, we can prove, as before, that *AB* is double *AF*, and *CD* double *CG*. And, since *AE* equals *CE*, the square on *AE* equals the square on *CE*. But the sum of the squares on *EF* and *FA* equals the square on *AE*, and the sum of the squares on *EG* and *GC* equals the square on *CE*.

I.47

Therefore the sum of the squares on *EF* and *FA* equals the sum of the squares on *EG* and *GC*, of which the square on *EF* equals the square on *EG*, for *EF* equals *EG*. Therefore the remaining square on *AF* equals the square on *CG*. Therefore *AF* equals *CG*. And *AB* is double *AF*, and *CD* double *CG*, therefore *AB* equals *CD*.

Therefore *equal straight lines in a circle are equally distant from the center, and those which are equally distant from the center equal one another*.

Q.E.D.

Proposition 15

Of straight lines in a circle the diameter is greatest, and of the rest the nearer to the center is always greater than the more remote.

Let *ABCD* be a circle, *AD* its diameter, and *E* its center. Let *BC* be nearer to the center *AD*, and *FG* more remote.
I say that *AD* is greatest and *BC* greater than *FG*.

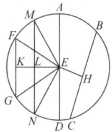

Draw *EH* and *EK* from the center *E* perpendicular to *BC* and *FG*.

I.12

Then, since *BC* is nearer to the center and *FG* more remote, *EK* is greater than *EH*.

III.Def.5

Make *EL* equal to *EH*. Draw *LM* through *L* at right angles to *EK*, and carry it through to *N*. Join *ME*, *EN*, *FE*, and *EG*.

I.3
I.11

Then, since *EH* equals *EL*, *BC* also equals *MN*. III.14

Again, since *AE* equals *EM*, and *ED* equals *EN*, *AD* equals the sum of *ME* and *EN*.

But the sum of *ME* and *EN* is greater than *MN*, and *MN* equals *BC*, therefore *AD* is greater than *BC*. I.20

And, since the two sides *ME* and *EN* equal the two sides *FE* and *EG*, and the angle *MEN* greater than the angle *FEG*, therefore the base *MN* is greater than the base *FG*. I.24

But *MN* was proved equal to *BC*.

Therefore the diameter *AD* is greatest and *BC* greater than *FG*.

Therefore *of straight lines in a circle the diameter is greatest, and of the rest the nearer to the center is always greater than the more remote.*

Q.E.D.

Proposition 16

The straight line drawn at right angles to the diameter of a circle from its end will fall outside the circle, and into the space between the straight line and the circumference another straight line cannot be interposed, further the angle of the semicircle is greater, and the remaining angle less, than any acute rectilinear angle.

Let *ABC* be a circle about *D* as center and *AB* as diameter.
I say that the straight line drawn from *A* at right angles to *AB* from its end will fall outside the circle.

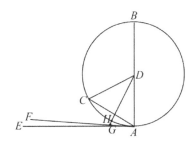

For suppose it does not, but, if possible, let it fall within as *CA*, and join *DC*.

Since *DA* equals *DC*, the angle *DAC* also equals the angle *ACD*. I.5

But the angle *DAC* is right, therefore the angle *ACD* is also right. Thus, in the triangle *ACD*, the two angles *DAC* and *ACD* equal two right angles, which is impossible. I.17

Therefore the straight line drawn from the point *A* at right angles to *BA* will not fall within the circle.

Similarly we can prove that neither will it fall on the circumference, therefore it will fall outside.

Let it fall as *AE*.

I say next that into the space between the straight line *AE* and the circumference *CHA* another straight line cannot be interposed.

For, if possible, let another straight line be so interposed, as *FA*. Draw *DG* from the point *D* perpendicular to *FA*.   I.12

Then, since the angle *AGD* is right, and the angle *DAG* is less than a right angle, *AD* is greater than *DG*.   I.17
  I.19

But *DA* equals *DH*, therefore *DH* is greater than *DG*, the less greater than the greater, which is impossible.

Therefore another straight line cannot be interposed into the space between the straight line and the circumference.

I say further that the angle of the semicircle contained by the straight line *BA* and the circumference *CHA* is greater than any acute rectilinear angle, and the remaining angle contained by the circumference *CHA* and the straight line *AE* is less than any acute rectilinear angle.

For, if there is any rectilinear angle greater than the angle contained by the straight line *BA* and the circumference *CHA*, and any rectilinear angle less than the angle contained by the circumference *CHA* and the straight line *AE*, then into the space between the circumference and the straight line *AE* a straight line will be interposed such as will make an angle contained by straight lines which is greater than the angle contained by the straight line *BA* and the circumference CHA, and another angle contained by straight lines which is less than the angle contained by the circumference CHA and the straight line *AE*.

But such a straight line cannot be interposed, therefore there will not be any acute angle contained by straight lines which is greater than the angle contained by the straight line *BA* and the circumference *CHA*, nor yet any acute angle contained by straight lines which is less than the angle contained by the circumference *CHA* and the straight line *AE*.

Therefore *the straight line drawn at right angles to the diameter of a circle from its end will fall outside the circle, and into the space between the straight line and the circumference another straight line cannot be interposed, further the angle of the semicircle is greater, and the remaining angle less, than any acute rectilinear angle.*

Q.E.D.

Corollary

From this it is manifest that *the straight line drawn at right angles to the diameter of a circle from its extremity touches the circle.*

Proposition 17

From a given point to draw a straight line touching a given circle.

Let *A* be the given point, and *BCD* the given circle.

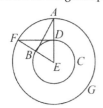

It is required to draw from the point *A* a straight line touching the circle *BCD*.

Take the center *E* of the circle, and join *AE*. Describe the circle *AFG* with center *E* and radius *EA*. Draw *DF* from *D* at right angles to *EA*. Join *EF* and *AB*.

III.1
I.11

I say that *AB* has been drawn from the point *A* touching the circle *BCD*.

For, since *E* is the center of the circles *BCD* and *AFG*, *EA* equals *EF*, and *ED* equals *EB*. Therefore the two sides *AE* and *EB* equal the two sides *FE* and *ED*, and they contain a common angle, the angle at *E*, therefore the base *DF* equals the base *AB*, and the triangle *DEF* equals the triangle *BEA*, and the remaining angles to the remaining angles, therefore the angle *EDF* equals the angle *EBA*.

I.4

But the angle *EDF* is right, therefore the angle *EBA* is also right.

Now *EB* is a radius, and the straight line drawn at right angles to the diameter of a circle, from its end, touches the circle, therefore *AB* touches the circle *BCD*.

III.16,Cor

Therefore from the given point *A* the straight line *AB* has been drawn touching the circle *BCD*.

Q.E.F.

Proposition 18

If a straight line touches a circle, and a straight line is joined from the center to the point of contact, the straight line so joined will be perpendicular to the tangent.

For let a straight line *DE* touch the circle *ABC* at the point *C*. Take the center *F* of the circle *ABC*, and join *FC* from *F* to *C*.

III.1

I say that *FC* is perpendicular to *DE*.

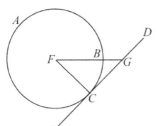

For, if not, draw *FG* from *F* perpendicular to *DE*.

I.12

Then, since the angle *FGC* is right, the angle *FCG* is acute, and the side opposite the greater angle is greater, therefore *FC* is greater than *FG*.

I.17
I.19

But *FC* equals *FB*, therefore *FB* is also greater than *FG*, the less greater than the greater, which is impossible.

Therefore *FG* is not perpendicular to *DE*.

Similarly we can prove that neither is any other straight line except *FC*. Therefore *FC* is perpendicular to *DE*.

Therefore *if a straight line touches a circle, and a straight line is joined from the center to the point of contact, the straight line so joined will be perpendicular to the tangent.*

Q.E.D.

Proposition 19

If a straight line touches a circle, and from the point of contact a straight line is drawn at right angles to the tangent, the center of the circle will be on the straight line so drawn.

For let a straight line DE touch the circle ABC at the point C. Draw CA from C at right angles to DE.

I.11

I say that the center of the circle is on AC.

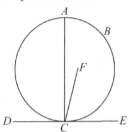

For suppose it is not, but, if possible, let F be the center, and join CF.

Since a straight line DE touches the circle ABC, and FC has been joined from the center to the point of contact, FC is perpendicular to DE; therefore the angle FCE is right.

III.18

But the angle ACE is also right, therefore the angle FCE equals the angle ACE, the less equals the greater, which is impossible.

Therefore F is not the center of the circle ABC.

Similarly we can prove that neither is any other point except a point on AC.

Therefore *if a straight line touches a circle, and from the point of contact a straight line is drawn at right angles to the tangent, the center of the circle will be on the straight line so drawn.*

Q.E.D.

Exercises on Propositions III.1 – III.19

1. Given: Circle ABCD with center P
   $BE = ED$
   Prove: $\angle APE > \angle EPC$

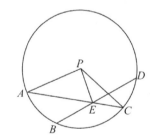

2. Given: Circles ABC and BDE touch at point B.
   $FA \parallel GD$
   Connect points FB, BG, AB, and BD.
   Prove: ABD is a straight line.

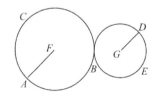

3. Given: Circle *ABCD* with center *P*
   $PE \perp AB$, $PF \perp CD$
   $BE = CF$
   Prove: *AB* and *CD* are equidistant from P

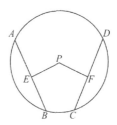

4. Given: Circle *ABCD* with center *E*
   $\angle AEB > \angle CED$
   Prove: AB is closer to E than CD

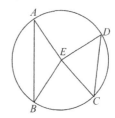

5. In circle *ABCD*, AC is a diameter and *P* is a point on the diameter, but not the center.

   Prove that the perpendicular to *AC* through *P* is the shortest line through P within the circle *ABCD*.

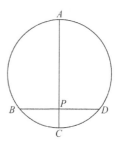

6. Given: Circle *ADBC* with diameter *CD*
   CD bisects *AB*
   ABCE is a parallelogram

   Prove: *EC* is tangent to circle *ABC*

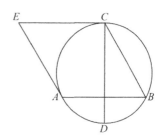

7. Equal circles *ABC* and *CDE*, with centers *F* and *G* respectively, touch each other at *C*. *PC* is also tangent to *ABC* at *C*.

   Prove *PC* is tangent to circle *CDE* at *C*. Then prove that *PF* = *PG*

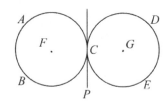

8. Given: Circles *ABC* and *DEF* with a common center *G*.
           DE is tangent to *ABC* at *B*.
   Prove: *B* bisects *DE*

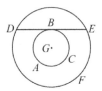

Proposition 20

In a circle the angle at the center is double the angle at the circumference when the angles have the same circumference as base.

Let *ABC* be a circle, let the angle *BEC* be an angle at its center, and the angle *BAC* an angle at the circumference, and let them have the same circumference *BC* as base.
I say that the angle *BEC* is double the angle *BAC*.

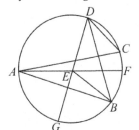

Join *AE*, and draw it through to *F*.

Then, since *EA* equals *EB*, the angle *EAB* also equals the angle *EBA*. Therefore the sum of the angles *EAB* and *EBA* is double the angle *EAB*.   I.5

But the angle *BEF* equals the sum of the angles *EAB* and *EBA*, therefore the angle *BEF*, is also double the angle *EAB*.   I.32.

For the same reason the angle *FEC* is also double the angle *EAC*.
Therefore the whole angle *BEC* is double the whole angle *BAC*.
Again let another straight line be inflected, and let there be another angle *BDC*. Join *DE* and produced it to *G*.
Similarly then we can prove that the angle *GEC* is double the angle *EDC*, of which the angle *GEB* is double the angle *EDB*. Therefore the remaining angle *BEC* is double the angle *BDC*.

Therefore *in a circle the angle at the center is double the angle at the circumference when the angles have the same circumference as base.*

                                                                                                                               Q.E.D.

Proposition 21

In a circle the angles in the same segment equal one another.

Let $ABCD$ be a circle, and let the angles BAD and BED be angles in the same segment $BAED$.

I say that the angles BAD and BED equal one another.

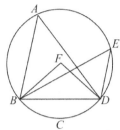

Take the center F of the circle $ABCD$, and join BF and FD.

Now, since the angle BFD is at the center, and the angle BAD at the circumference, and they have the same circumference BCD as base, therefore the angle BFD is double the angle BAD.   III.20

For the same reason the angle BFD is also double the angle BED. Therefore the angle BAD equals the angle BED.

Therefore *in a circle the angles in the same segment equal one another.*

Q.E.D

Proposition 22

The sum of the opposite angles of quadrilaterals in circles equals two right angles.

Let $ABCD$ be a circle, and let $ABCD$ be a quadrilateral in it.

I say that the sum of the opposite angles equals two right angles.

Join AC and BD.

Then, since in any triangle the sum of the three angles equals two right angles, the sum of the three angles CAB, ABC, and BCA of the triangle ABC equals two right angles.   I.32

But the angle CAB equals the angle BDC, for they are in the same segment $BADC$, and the angle ACB equals the angle ADB, for they are in the same segment $ADCB$, therefore the whole angle ADC equals the sum of the angles BAC and ACB.   III.21

Add the angle ABC to each. Therefore the sum of the angles ABC, BAC, and ACB equals the sum of the angles ABC and ADC. But the sum of the angles ABC, BAC, and ACB equals two right angles, therefore the sum of the angles ABC and ADC also equal two right angles.

Similarly we can prove that the sum of the angles BAD and DCB also equals two right angles.

Therefore *the sum of the opposite angles of quadrilaterals in circles equals two right angles.*

Q.E.D.

Proposition 23

On the same straight line there cannot be constructed two similar and unequal segments of circles on the same side.

For, if possible, on the same straight line AB let two similar and unequal segments of circles ACB and ADB be constructed on the same side. Draw ACD through, and join CB and DB.

Then, since the segment ACB is similar to the segment ADB, and similar segments of circles are those which admit equal angles, the angle ACB equals the angle ADB, the exterior to the interior, which is impossible.

III Def.11
I.16

Therefore *on the same straight line there cannot be constructed two similar and unequal segments of circles on the same side.*

Q.E.D.

Proposition 24

Similar segments of circles on equal straight lines equal one another.

Let AEB and CFD be similar segments of circles on equal straight lines AB and CD.

I say that the segment AEB equals the segment CFD.

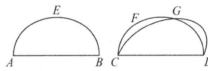

For, if the segment AEB is superposed on CFD, and if the point A is placed on C and the straight line AB on CD, then the point B coincides with the point D, because AB equals CD, and, AB coinciding with CD, the segment AEB also coincides with CFD.

III.23

For, if the straight line AB coincides with CD but the segment AEB does not coincide with CFD, then it either falls within it, or outside it, or it falls awry, as CGD, and a circle cuts a circle at more points than two, which is impossible.

III.10

Therefore, if the straight line AB is superposed on CD, then the segment AEB does not fail to coincide with CFD also, therefore it coincides with it and equals it.

Therefore *similar segments of circles on equal straight lines equal one another.*

Q.E.D.

Proposition 25

Given a segment of a circle, to describe the complete circle of which it is a segment.

Let ABC be the given segment of a circle.
It is required to describe the complete circle belonging to the segment ABC, that is, of which it is a segment.

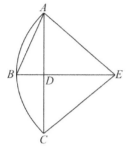

Bisect AC at D, draw DB from the point D at right angles to AC, and join AB.	I.10
The angle ABD is then greater than, equal to, or less than the angle BAD.	I.11
First let it be greater. Construct the angle BAE on the straight line BA, and at the point A on it, equal to the angle ABD. Draw DB through to E, and join EC.	I.23
Then, since the angle ABE equals the angle BAE, the straight line EB also equals EA.	I.6

And, since AD equals DC, and DE is common, the two sides AD and DE equal the two sides CD and DE respectively, and the angle ADE equals the angle CDE, for each is right, therefore the base AE equals the base CE.   I.4

But AE was proved equal to BE, therefore be also equals CE. Therefore the three straight lines AE, EB, and EC equal one another.

Therefore the circle drawn with center E and radius one of the straight lines AE, EB, or EC also passes through the remaining points and has been completed.   III.9

Therefore, given a segment of a circle, the complete circle has been described.
And it is manifest that the segment ABC is less than a semicircle, because the center E happens to be outside it.

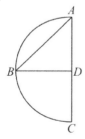

Similarly, even if the angle ABD equals the angle BAD and AD being equal to each of the two BD and DC, the three straight lines DA, DB, and DC will equal one another, D will be the center of the completed circle, and ABC will clearly be a semicircle.

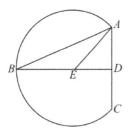

But, if the angle *ABD* is less than the angle *BAD*, and if we construct, on the straight line *BA* and at the point *A* on it, an angle equal to the angle *ABD*, the center will fall on *DB* within the segment *ABC*, and the segment *ABC* will clearly be greater than a semicircle.

I.23

Therefore, given a segment of a circle, the complete circle has been described.

Q.E.F.

Proposition 26

In equal circles equal angles stand on equal circumferences whether they stand at the centers or at the circumferences.

Let *ABC* and *DEF* be equal circles, and in them let there be equal angles, namely at the centers the angles *BGC* and *EHF*, and at the circumferences the angles *BAC* and *EDF*.

I say that the circumference *BKC* equals the circumference *ELF*.

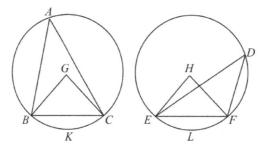

Join *BC* and *EF*.

Now, since the circles *ABC* and *DEF* are equal, the radii are equal.

Thus the two straight lines *BG* and *GC* equal the two straight lines *EH* and *HF*, and the angle at *G* equals the angle at *H*, therefore the base *BC* equals the base *EF*.

I.4

And, since the angle at *A* equals the angle at *D*, the segment *BAC* is similar to the segment *EDF*, and they are upon equal straight lines.

III.Def.11

But similar segments of circles on equal straight lines equal one another, therefore the segment *BAC* equals *EDF*. But the whole circle *ABC* also equals the whole circle *DEF*, therefore the remaining circumference *BKC* equals the circumference *ELF*.

III.24

Therefore *in equal circles equal angles stand on equal circumferences whether they stand at the centers or at the circumferences.*

Q.E.D.

Proposition 27

In equal circles angles standing on equal circumferences equal one another whether they stand at the centers or at the circumferences.

For in equal circles ABC and DEF, on equal circumferences BC and EF, let the angles BGC and EHF stand at the centers G and H, and the angles BAC and EDF at the circumferences.

I say that the angle BGC equals the angle EHF, and the angle BAC equals the angle EDF.

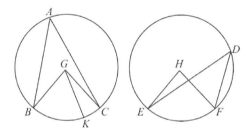

For, if the angle BGC does not equal the angle EHF, one of them is greater. Let the angle BGC be greater. Construct the angle BGK equal to the angle EHF on the straight line BG and at the point G on it.   I.23

Now equal angles stand on equal circumferences when they are at the centers, therefore the circumference BK equals the circumference EF.   III.26

But EF equals BC, therefore BK also equals BC, the less equals the greater, which is impossible.

Therefore the angle BGC is not unequal to the angle EHF, therefore it equals it.

And the angle at A is half of the angle BGC, and the angle at D half of the angle EHF, therefore the angle at A also equals the angle at D.   III.20

Therefore *in equal circles angles standing on equal circumferences equal one another whether they stand at the centers or at the circumferences.*

Q.E.D.

Proposition 28

In equal circles equal straight lines cut off equal circumferences, the greater circumference equals the greater and the less equals the less.

Let ABC and DEF be equal circles, and in the circles let AB and DE be equal straight lines cutting off ACB and DFE as greater circumferences and AGB and DHE as lesser. I say that the greater circumference ACB equals the greater circumference DFE, and the less circumference AGB equals DHE.

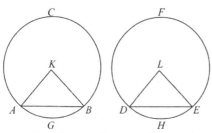

Take the centers *K* and *L* of the circles, and join *AK*, *KB*, *DL*, and *LE*.	III.1
Now, since the circles are equal, the radii are also equal, therefore the two sides *AK* and *KB* equal the two sides *DL* and *LE*, and the base *AB* equals the base *DE*, therefore the angle *AKB* equals the angle *DLE*.	I.8

But equal angles stand on equal circumferences when they are at the centers, therefore the circumference *AGB* equals *DHE*.   III.26

And the whole circle *ABC* also equals the whole circle *DEF*, therefore the remaining circumference *ACB* also equals the remaining circumference *DFE*.

Therefore *in equal circles equal straight lines cut off equal circumferences, the greater circumference equals the greater and the less equals the less.*

Q.E.D.

Proposition 29

In equal circles straight lines that cut off equal circumferences are equal.

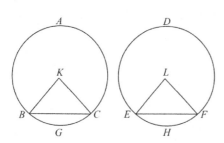

Let *ABC* and *DEF* be equal circles, and in them let equal circumferences *BGC* and *EHF* be cut off. Join the straight lines *BC* and *EF*.

I say that *BC* equals *EF*.

Take the centers *K* and *L* of the circles. Join *BK*, *KC*, *EL*, and *LF*.	III.1
Now, since the circumference *BGC* equals the circumference *EHF*, the angle *BKC* also equals the angle *ELF*.	III.27
And, since the circles *ABC* and *DEF* are equal, the radii are also equal, therefore the two sides *BK* and *KC* equal the two sides *EL* and *LF*, and they contain equal angles, therefore the base *BC* equals the base *EF*.	I.4

Therefore *in equal circles straight lines that cut off equal circumferences are equal.*

Q.E.D.

Proposition 30

To bisect a given circumference.

Let *ADB* be the given circumference.
It is required to bisect the circumference *ADB*.

Join *AB*, and bisect it at *C*. Draw *CD* from the point *C* at right angles to the straight line *AB*. Join *AD* and *DB*.   I.10
I.11

Then, since *AC* equals *CB*, and *CD* is common, the two sides *AC* and *CD* equal the two sides *BC* and *CD*, and the angle *ACD* equals the angle *BCD*, for each is right, therefore the base *AD* equals the base *DB*.   I.4

But equal straight lines cut off equal circumferences, the greater equal to the greater, and the less to the less, and each of the circumferences *AD* and *DB* is less than a semicircle, therefore the circumference *AD* equals the circumference *DB*.   III.28

Therefore the given circumference has been bisected at the point *D*.

Q.E.F.

Proposition 31

In a circle the angle in the semicircle is right, that in a greater segment less than a right angle, and that in a less segment greater than a right angle; further the angle of the greater segment is greater than a right angle, and the angle of the less segment is less than a right angle.

Let *ABCD* be a circle, let *BC* be its diameter, and *E* its center. Join *BA*, *AC*, *AD*, and *DC*. I say that the angle *BAC* in the semicircle *BAC* is right, the angle *ABC* in the segment *ABC* greater than the semicircle is less than a right angle, and the angle *ADC* in the segment *ADC* less than the semicircle is greater than a right angle.

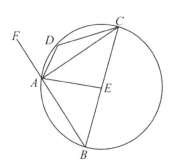

Join *AE*, and carry *BA* through to *F*.

Then, since *BE* equals *EA*, the angle *ABE* also equals the angle *BAE*. Again, since *CE* equals *EA*, the angle *ACE* also equals the angle *CAE*. Therefore the whole angle *BAC* equals the sum of the two angles *ABC* and *ACB*.   I.5

But the angle *FAC* exterior to the triangle *ABC* also equals the sum of the two angles *ABC* and *ACB*. Therefore the angle *BAC* also equals the angle *FAC*. Therefore each is right. Therefore the angle *BAC* in the semicircle *BAC* is right.   I.32

Next, since in the triangle *ABC* the sum of the two angles *ABC* and *BAC* is less than two right angles, and the angle *BAC* is a right angle, the angle *ABC* is less than a right angle. And it is the angle in the segment *ABC* greater than the semicircle.   I.17

Next, since *ABCD* is a quadrilateral in a circle, and the sum of the opposite angles of quadrilaterals in circles equals two right angles, while the angle *ABC* is less than a right angle, therefore the remaining angle *ADC* is greater than a right angle. And it is the angle in the segment *ADC* less than the semicircle.   III.22

I say further that the angle of the greater segment, namely that contained by the circumference *ABC* and the straight line *AC*, is greater than a right angle, and the angle of the less segment, namely that contained by the circumference *ADC* and the straight line *AC*, is less than a right angle.

This is at once manifest. For, since the angle contained by the straight lines *BA* and *AC* is right, the angle contained by the circumference *ABC* and the straight line *AC* is greater than a right angle.

Again, since the angle contained by the straight lines *AC* and *AF* is right, the angle contained by the straight line *CA* and the circumference *ADC* is less than a right angle.

Therefore *in a circle the angle in the semicircle is right, that in a greater segment less than a right angle, and that in a less segment greater than a right angle; further the angle of the greater segment is greater than a right angle, and the angle of the less segment is less than a right angle.*

Q.E.D.

Proposition 32

If a straight line touches a circle, and from the point of contact there is drawn across, in the circle, a straight line cutting the circle, then the angles which it makes with the tangent equal the angles in the alternate segments of the circle.

For let a straight line *EF* touch the circle *ABCD* at the point *B*, and from the point *B* let there be drawn across, in the circle *ABCD*, a straight line *BD* cutting it.
I say that the angles which *BD* makes with the tangent *EF* equal the angles in the alternate segments of the circle, that is, that the angle *FBD* equals the angle constructed in the segment *BAD*, and the angle *EBD* equals the angle constructed in the segment *DCB*.

Draw *BA* from *B* at right angles to *EF*, take a point *C* at random on the circumference *BD*, and join *AD*, *DC*, and *CB*.   I.11

Then, since a straight line *EF* touches the circle *ABCD* at *B*, and *BA* has been drawn from the point of contact at right angles to the tangent, the center of the circle *ABCD* is on *BA*.   III.19

Therefore *BA* is a diameter of the circle *ABCD*. Therefore the angle *ADB*, being an angle in a semicircle, is right.   III.31

Therefore the sum of the remaining angles *BAD* and *ABD* equals one right angle.   I.32

But the angle *ABF* is also right, therefore the angle *ABF* equals the sum of the angles *BAD* and *ABD*.

Subtract the angle *ABD* from each. Therefore the remaining angle *DBF* equals the angle *BAD* in the alternate segment of the circle.

Next, since *ABCD* is a quadrilateral in a circle, the sum of its opposite angles equals two right angles. III.22

But the sum of the angles *DBF* and *DBE* also equals two right angles, therefore the sum of the angles *DBF* and *DBE* equals the sum of the angles *BAD* and *BCD*, of which the angle *BAD* was proved equal to the angle *DBF*, therefore the remaining angle *DBE* equals the angle *DCB* in the alternate segment *DCB* of the circle.

Therefore *if a straight line touches a circle, and from the point of contact there is drawn across, in the circle, a straight line cutting the circle, then the angles which it makes with the tangent equal the angles in the alternate segments of the circle.*

Q.E.D.

Exercises on Propositions III.1 – III.32

1. In the drawing to the right *ABC* is a circle with center *D*.

    a. If *BDC* is a right angle, how big is angle *BAC*?

    b. If *BAC* is a right angle, what is the measure of angle *BDC*? Draw a picture of this case.

    c. Read Proposition III.31

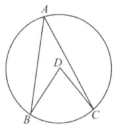

2. thout bisecting any angles, construct an angle that is ½ a right angle; ¼ of a right angle; 1/8 of a right angle.

3. Given: Circle *ABCD* with ∠*ACB* = ∠*DBC*

    Prove: △*ABC* and △*DCB* are equiangular, and the sides of △*ABC* are equal to the sides of △*DCB*, respectively

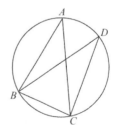

4. Circles ADCB and BGFE touch one another at point B. Lines AE and CG pass through point B.

   Prove: ∠ADC equals ∠GFE

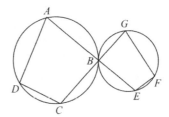

5. Given: Equal circles ABC and DEF

   Prove the following:

   a. if ∠BAC = ∠EDF then line BC = line EF
   b. if line BC = line EF then ∠BAC = ∠EDF

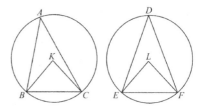

6. In the drawing to the right, quadrilateral ABCD has been inscribed in circle ABCD. If ∠ABC is greater than a right angle, prove that circumference (arc) ABC is less than a semicircle.

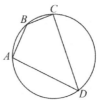

7. Given: Line CD tangent to circle ABC at C
   AB ∥ CD
   Prove: △ABC is isosceles

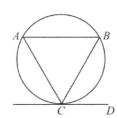

8. In the drawing to the right, lines AP and BP touch circle ABC at A and B respectively. Without altering the drawing, prove AP = BP.

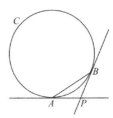

9. Given: Circle ABC with center P and triangle ABC inscribed in it.

   Express ∠PBC in terms of angles x, y, and any constants you need (like, for example, a certain number of right angles).

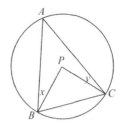

Proposition 33

On a given straight line to describe a segment of a circle admitting an angle equal to a given rectilinear angle.

Let AB be the given straight line, and the angle at C the given rectilinear angle. It is required to describe on the given straight line AB a segment of a circle admitting an angle equal to the angle at C.
The angle at C is then acute, or right, or obtuse.

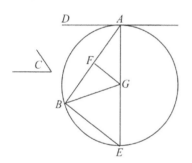

First let it be acute as in the first figure. Construct the angle BAD equal to the angle at C on the straight line AB and at the point A. Therefore the angle BAD is also acute.

I.23

Draw AE at right angles to DA. Bisect AB at F. Draw FG from the point F at right angles to AB, and join GB.

I.10
I.12

Then, since AF equals FB, and FG is common, the two sides AF and FG equal the two sides BF and FG, and the angle AFG equals the angle BFG, therefore the base AG equals the base BG.

I.4

Therefore the circle described with center G and radius GA passes through B also.
Draw the circle as ABE, and join EB.

Now, since AD is drawn from A, the end of the diameter AE, at right angles to AE, therefore AD touches the circle ABE.

III.16, Cor.

Since then a straight line AD touches the circle ABE, and from the point of contact at A a straight line AB has been drawn across in the circle ABE, the angle DAB equals the angle AEB in the alternate segment of the circle.

III.32

But the angle *DAB* equals the angle at *C*, therefore the angle at *C* also equals the angle *AEB*.

Therefore on the given straight line *AB* the segment *AEB* of a circle has been described admitting the angle *AEB* equal to the given angle, the angle at *C*.

Next let the angle at *C* be right, and let it be again required to describe on *AB* a segment of a circle admitting an angle equal to the right angle at *C*.

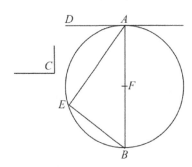

Let the angle *BAD* be constructed equal to the right angle at *C*, as is the case in the second figure. Bisect *AB* at *F*. Describe the circle *AEB* with center *F* and radius either *FA* or *FB*.

I.23
I.10

Therefore the straight line *AD* touches the circle *ABE*, because the angle at A is right.

III.16, Cor.

And the angle *BAD* equals the angle in the segment *AEB*, for the latter too is itself a right angle, being an angle in a semicircle.

III.31

But the angle *BAD* also equals the angle at *C*, therefore the angle *AEB* also equals the angle at *C*.

Therefore again the segment *AEB* of a circle has been described on *AB* admitting an angle equal to the angle at *C*.

Next, let the angle at *C* be obtuse.

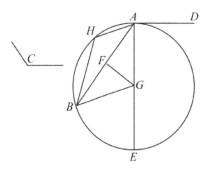

Construct the angle *BAD* equal to *C* on the straight line *AB* and at the point A as is the case in the third figure. Draw *AE* at right angles to *AD*. Bisect *AB* again at *F*. Draw *FG* at right angles to *AB*, and join *GB*.

I.23
I.11
I.12

Then, since *AF* again equals *FB*, and *FG* is common, the two sides *AF* and *FG* equal the two sides *BF* and *FG*, and the angle *AFG* equals the angle *BFG*, therefore the base *AG* equals the base *BG*.

I.4

Therefore the circle described with center *G* and radius *GA* also passes through B. Let it so pass, as *AEB*.

Now, since *AD* is drawn at right angles to the diameter *AE* from its end, *AD* touches the circle *AEB*.

III.16, Cor.

And *AB* has been drawn across from the point of contact at A, therefore the angle *BAD* equals the angle constructed in the alternate segment *AHB* of the circle.

III.32

But the angle BAD equals the angle at C.
Therefore the angle in the segment AHB also equals the angle at C.
Therefore on the given straight line AB the segment AHB of a circle has been described admitting an angle equal to the angle at C.

Q.E.F.

Proposition 34

From a given circle to cut off a segment admitting an angle equal to a given rectilinear angle.

Let ABC be the given circle, and the angle at D the given rectilinear angle.
It is required to cut off from the circle ABC a segment admitting an angle equal to the given rectilinear angle, the angle at D.

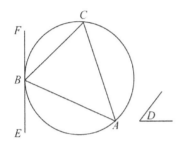

Draw EF touching ABC at the point B. Construct the angle FBC equal to the angle at D on the straight line FB and at the point B on it.

III.17
I.23

Then, since a straight line EF touches the circle ABC, and BC has been drawn across from the point of contact at B, the angle FBC equals the angle constructed in the alternate segment BAC.

III.32

But the angle FBC equals the angle at D, therefore the angle in the segment BAC equals the angle at D.

Therefore from the given circle ABC the segment BAC has been cut off admitting an angle equal to the given rectilinear angle, the angle at D.

Q.E.F.

For the exercises on this page you will need to know the following definitions.

Definitions

1. A rectilinear figure is said to be *inscribed in a circle* when each angle of the inscribed figure lies on the circumference of the circle.

2. A rectilinear figure is said to be *circumscribed about a circle* when each side of the circumscribed figure touches the circumference of the circle.

3. Similarly a circle is said to be *inscribed in a figure* when the circumference of the circle touches each side of the figure in which it is inscribed.

4. A circle is said to be *circumscribed about a figure* when the circumference of the circle passes through each angle of the figure about which it is circumscribed.

Exercises on Books I – III

These exercises are constructions. For each exercise, figure out how to do the required construction, then write up a proof justifying your construction. If you need to, refer back to earlier construction propositions for proper form.

1. In a given circle, inscribe a square.

2. About a given circle, circumscribe a square.

3. In a given square, inscribe a circle.

4. About a given square, circumscribe a circle.

Book IV

Definitions

1. A rectilinear figure is said to be *inscribed in a rectilinear figure* when the respective angles of the inscribed figure lie on the respective sides of that in which it is inscribed.

2. Similarly a figure is said to be *circumscribed about a figure* when the respective sides of the circumscribed figure pass through the respective angles of that about which it is circumscribed.

3. A rectilinear figure is said to be *inscribed in a circle* when each angle of the inscribed figure lies on the circumference of the circle.

4. A rectilinear figure is said to be *circumscribed about a circle* when each side of the circumscribed figure touches the circumference of the circle.

5. Similarly a circle is said to be *inscribed in a figure* when the circumference of the circle touches each side of the figure in which it is inscribed.

6. A circle is said to be *circumscribed about a figure* when the circumference of the circle passes through each angle of the figure about which it is circumscribed.

7. A straight line is said to be *fitted into a circle* when its ends are on the circumference of the circle.

Book IV Propositions

Proposition 1

To fit a straight line into a given circle equal to a given straight line which is not greater than the diameter of the circle.

Let ABC be the given circle, and D the given straight line not greater than the diameter of the circle.

It is required to fit a straight line into the circle ABC equal to the straight line D.

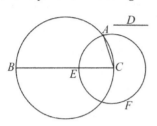

Draw a diameter BC of the circle ABC.

If BC equals D, then that which was proposed is done, for BC has been fitted into the circle ABC equal to the straight line D.\

But, if BC is greater than D, make CE equal to D, describe the circle EAF with center C and radius CE, and join CA.

I.3

Then, since the point C is the center of the circle EAF, CA equals CE.

But CE equals D, therefore D also equals CA.

Therefore CA has been fitted into the given circle ABC equal to the given straight line D.

IV.Def.7

Q.E.F.

Proposition 2

To inscribe a triangle equiangular with a given triangle in a given circle.

Let ABC be the given circle, and DEF the given triangle.

It is required to inscribe a triangle equiangular with the triangle DEF in the circle ABC.

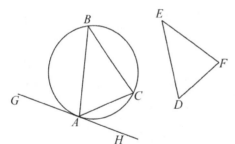

Draw GH touching the circle ABC at A. Construct the angle HAC equal to the angle DEF on the straight line AH and at the point A on it, and construct the angle GAB equal to the angle DFE on the straight line AG and at the point A on it. Join BC.

III.16, Cor
I.23

Then, since a straight line AH touches the circle ABC, and from the point of contact at A the straight line AC is drawn across in the circle, therefore the angle HAC equals the angle ABC in the alternate segment of the circle.

III.32

But the angle *HAC* equals the angle *DEF*, therefore the angle *ABC* also equals the angle *DEF*.

For the same reason the angle *ACB* also equals the angle *DFE*, therefore the remaining angle *BAC* also equals the remaining angle *EDF*. I.32

Therefore a triangle equiangular with the given triangle has been inscribed in the given circle. IV.Def.2

Q.E.F.

Proposition 3

To circumscribe a triangle equiangular with a given triangle about a given circle.

Let *ABC* be the given circle, and *DEF* the given triangle.

It is required to circumscribe a triangle equiangular with the triangle *DEF* about the circle *ABC*.

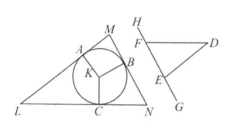

Produce *EF* in both directions to the points *G* and *H*. Take the center *K* of the circle *ABC*, and draw a radius *KB* at random. On the straight line *KB* and at the point *K* on it, construct the angle *BKA* equal to the angle *DEG*, and the angle *BKC* equal to the angle *DFH*. Through the points *A*, *B*, and *C* draw *LAM*, *MBN*, and *NCL* touching the circle *ABC*.

III.1
I.23
III.16, Cor

Now, since *LM*, *MN*, and *NL* touch the circle *ABC* at the points *A*, *B*, and *C*, and *KA*, *KB*, and *KC* have been joined from the center *K* to the points *A*, *B*, and *C*, therefore the angles at the points *A*, *B*, and *C* are right. III.18

And, since the four angles of the quadrilateral *AMBK* equal four right angles, inasmuch as *AMBK* is in fact divisible into two triangles, and the angles *KAM* and *KBM* are right, therefore the sum of the remaining angles *AKB* and *AMB* equals two right angles.

But the sum of the angles *DEG* and *DEF* also equals two right angles, therefore the sum of the angles *AKB* and *AMB* equals the sum of the angles *DEG* and *DEF*, of which the angle *AKB* equals the angle *DEG*, therefore the remaining angle *AMB* equals the remaining angle *DEF*. I.13

Similarly it can be proved that the angle *LNB* also equals the angle *DFE*, therefore the remaining angle *MLN* equals the angle *EDF*. I.32

Therefore the triangle *LMN* is equiangular with the triangle *DEF*, and it has been circumscribed about the circle *ABC*. IV.Def.4

Therefore a triangle equiangular with the given triangle has been circumscribed about a given circle.

Q.E.F.

Proposition 4

To inscribe a circle in a given triangle.
Let ABC be the given triangle.
It is required to inscribe a circle in the triangle ABC.

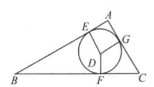

Bisect the angles ABC and ACB by the straight lines BD and CD, and let these meet one another at the point D. Draw DE, DF, and DG from D perpendicular to the straight lines AB, BC, and CA.

I.9
I.12

Now, since the angle ABD equals the angle CBD, and the right angle BED also equals the right angle BFD, EBD and FBD are two triangles having two angles equal to two angles and one side equal to one side, namely that opposite one of the equal angles, which is BD common to the triangles, therefore they will also have the remaining sides equal to the remaining sides, therefore DE equals DF.
For the same reason DG also equals DF.
Therefore the three straight lines DE, DF, and DG equal one another. Therefore the circle described with center D and radius one of the straight lines DE, DF, or DG also passes through the remaining points and touches the straight lines AB, BC, and CA, because the angles at the points E, F, and G are right.
For, if it cuts them, the straight line drawn at right angles to the diameter of the circle from its end will be found to fall within the circle, which was proved absurd, therefore the circle described with center D and radius one of the straight lines DE, DF, or DG does not cut the straight lines AB, BC, and CA Therefore it touches them, and is the circle inscribed in the triangle ABC.
Let it be inscribed as FGE.
Therefore the circle EFG has been inscribed in the given triangle ABC.

I.26

III.16
IV.Def.5

Q.E.F.

Proposition 5

To circumscribe a circle about a given triangle.
Let ABC be the given triangle.
It is required to circumscribe a circle about the given triangle ABC.

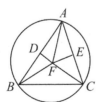 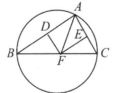 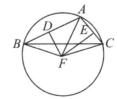

Bisect the straight lines AB and AC at the points D and E. Draw DF and EF from the points D and E at right angles to AB and AC. They will then meet within the triangle ABC, or on the straight line BC, or outside BC.   I.10
I.11

First let them meet within at F. Join FB, FC, and FA.

Then, since AD equals DB, and DF is common and at right angles, therefore the base AF equals the base FB.   I.4

Similarly we can prove that CF also equals AF, so that FB also equals FC, therefore the three straight lines FA and FB and FC equal one another.

Therefore the circle described with center F and radius one of the straight lines FA, FB, or FC also passes through the remaining points, and the circle is circumscribed about the triangle ABC.   IV.Def.6

Let it be circumscribed as ABC.

Next, let DF and EF meet on the straight line BC at F, as is the case in the second figure. Join AF.

Then, similarly, we can prove that the point F is the center of the circle circumscribed about the triangle ABC.

Next, let DF and EF meet outside the triangle ABC at F, as is the case in the third figure. Join AF, BF, and CF.

Then again, since AD equals DB, and DF is common and at right angles, therefore the base AF equals the base BF.   I.4

Similarly we can prove that CF also equals AF, so that BF also equals FC. Therefore the circle described with center F and radius one of the straight lines FA, FB, or FC also passes through the remaining points, and is circumscribed about the triangle ABC.   IV.Def.6

Therefore a circle has been circumscribed about the given triangle.

Q.E.F.

Corollary

And it is manifest that when the center of the circle falls within the triangle, the angle BAC, being in a segment greater than the semicircle, is less than a right angle, when the center falls on the straight line BC, the angle BAC, being in a semicircle, is right, and when the center of the circle falls outside the triangle, the angle BAC, being in a segment less than the semicircle, is greater than a right angle.   III.31

Proposition 6

To inscribe a square in a given circle.

Let ABCD be the given circle.
It is required to inscribe a square in the circle ABCD.

Draw two diameters AC and BD of the circle ABCD at right angles to one another, and join AB, BC, CD, and DA.   III.1
I.11

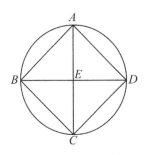

Then, since BE equals ED, for E is the center, and EA is common and at right angles, therefore the base AB equals the base AD. I.4

For the same reason each of the straight lines BC and CD also equals each of the straight lines AB and AD. Therefore the quadrilateral ABCD is equilateral.
I say next that it is also right-angled.

For, since the straight line BD is a diameter of the circle ABCD, therefore BAD is a semicircle, therefore the angle BAD is right. III.31

For the same reason each of the angles ABC, BCD, and CDA is also right. Therefore the quadrilateral ABCD is right-angled.

But it was also proved equilateral, therefore it is a square, and it has been inscribed in the circle ABCD.

Therefore the square ABCD has been inscribed in the given circle.

Q.E.F

Proposition 7

To circumscribe a square about a given circle.

Let ABCD be the given circle.
It is required to circumscribe a square about the circle ABCD.

Draw two diameters AC and BD of the circle ABCD at right angles to one another. III.1
Draw FG, GH, HK, and KF through the points A, B, C, and D touching the circle ABCD. I.11
III.16, Cor

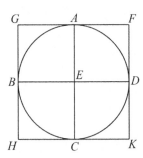

Then, since FG touches the circle ABCD, and EA has been joined from the center E to the point of contact at A, therefore the angles at A are right. III.18

For the same reason the angles at the points B, C, and D are also right.

Now, since the angle AEB is right, and the angle EBG is also right, therefore GH is parallel to AC. I.28

For the same reason AC is also parallel to FK, so that GH is also parallel to FK. I.30

Similarly we can prove that each of the straight lines GF and HK is parallel to BED.

Therefore GK, GC, AK, FB, and BK are parallelograms, therefore GF equals HK, and GH equals FK. I.34

And, since AC equals BD, and AC also equals each of the straight lines GH and FK, and BD equals each of the straight lines GF and HK, therefore the quadrilateral FGHK is equilateral.

I.34

I say next that it is also right-angled.

For, since GBEA is a parallelogram, and the angle AEB is right, therefore the angle AGB is also right.

I.34

Similarly we can prove that the angles at H, K, and F are also right.
Therefore FGHK is right-angled.
But it was also proved equilateral, therefore it is a square, and it has been circumscribed about the circle ABCD.
Therefore a square has been circumscribed about the given circle.

Q.E.F.

Proposition 8

To inscribe a circle in a given square.

Let ABCD be the given square.
It is required to inscribe a circle in the given square ABCD.

Bisect the straight lines AD and AB at the points E and F respectively. Draw EH through E parallel to either AB or CD, and draw FK through F parallel to either AD or BC. Therefore each of the figures AK, KB, AH, HD, AG, GC, BG, and GD is a parallelogram, and their opposite sides are evidently equal.

I.10
I.31
I.34

Now, since AD equals AB, and AE is half of AD, and AF half of AB, therefore AE equals AF, so that the opposite sides are also equal, therefore FG equals GE. Similarly we can prove that each of the straight lines GH and GK equals each of the straight lines FG and GE. Therefore the four straight lines GE, GF, GH, and GK equal one another.

Therefore the circle described with center G and radius one of the straight lines GE, GF, GH, or GK also passes through the remaining points.

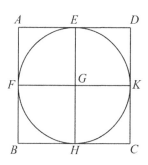

And it touches the straight lines AB, BC, CD, and DA, because the angles at E, F, H, and K are right.

For, if the circle cuts AB, BC, CD, or DA, the straight line drawn at right angles to the diameter of the circle from its end will fall within the circle, which was proved absurd. Therefore the circle described with center G and radius one of the straight lines GE, GF, GH, or GK does not cut the straight lines AB, BC, CD, and DA.

III.16

Therefore it touches them, and has been inscribed in the square ABCD.
Therefore a circle has been inscribed in the given square.

Q.E.F.

Proposition 9

To circumscribe a circle about a given square.

Let *ABCD* be the given square.
It is required to circumscribe a circle about the square *ABCD*.

Join *AC* and *BD*, and let them cut one another at *E*.

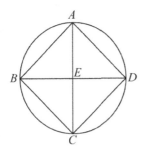

Then, since *DA* equals *AB*, and *AC* is common, therefore the two sides *DA* and *AC* equal the two sides *BA* and *AC*, and the base *DC* equals the base *BC*, therefore the angle *DAC* equals the angle *BAC*. I.8

Therefore the angle *DAB* is bisected by *AC*.
Similarly we can prove that each of the angles *ABC*, *BCD*, and *CDA* is bisected by the straight lines *AC* and *DB*.

Now, since the angle *DAB* equals the angle *ABC*, and the angle *EAB* is half of the angle *DAB*, and the angle *EBA* half of the angle *ABC*, therefore the angle *EAB* also equals the angle *EBA*, so that the side *EA* also equals *EB*. I.6

Similarly we can prove that each of the straight lines *EA* and *EB* equals each of the straight lines *EC* and *ED*.
Therefore the four straight lines *EA*, *EB*, *EC*, and *ED* equal one another.
Therefore the circle described with center *E* and radius one of the straight lines *EA*, *EB*, *EC*, or *ED* also passes through the remaining points, and it is circumscribed about the square *ABCD*.
Let it be circumscribed, as *ABCD*.
Therefore a circle has been circumscribed about the given square.

Q.E.F.

Exercises on Book IV

1. For octagons, state four propositions corresponding to Propositions IV.6 – IV.9. Prove one of them.

2. (For those who have used Appendix A to complete Book IV.) For 10-sided regular polygons, state four propositions corresponding to those for the polygons of Book IV and prove one of them.

3. Try to do Exercise 1 for 9-sided figures. Is there a problem? If so, what goes wrong?

Book V

Definitions

1. A magnitude is a *part* of a magnitude, the less of the greater, when it measures the greater.

2. The greater is a *multiple* of the less when it is measured by the less.

3. A *ratio* is a sort of relation in respect of size between two magnitudes of the same kind.

4. Magnitudes are said to *have a ratio* to one another which can, when multiplied, exceed one another.

5. Magnitudes are said to be *in the same ratio*, the first to the second and the third to the fourth, when, if any equimultiples whatever are taken of the first and third, and any equimultiples whatever of the second and fourth, the former equimultiples alike exceed, are alike equal to, or alike fall short of, the latter equimultiples respectively taken in corresponding order.

6. Let magnitudes which have the same ratio be called *proportional*.

7. When, of the equimultiples, the multiple of the first magnitude exceeds the multiple of the second, but the multiple of the third does not exceed the multiple of the fourth, then the first is said to have a *greater ratio* to the second than the third has to the fourth.

8. A proportion in three terms is the least possible.

9. When three magnitudes are proportional, the first is said to have to the third the *duplicate ratio* of that which it has to the second.

10. When four magnitudes are continuously proportional, the first is said to have to the fourth the *triplicate ratio* of that which it has to the second, and so on continually, whatever be the proportion.

11. Antecedents are said to *correspond* to antecedents, and consequents to consequents.

12. *Alternate ratio* means taking the antecedent in relation to the antecedent and the consequent in relation to the consequent.

13. *Inverse ratio* means taking the consequent as antecedent in relation to the antecedent as consequent.

14. A ratio *taken jointly* means taking the antecedent together with the consequent as one in relation to the consequent by itself.

15. A ratio *taken separately* means taking the excess by which the antecedent exceeds the consequent in relation to the consequent by itself.

16. *Conversion of a ratio* means taking the antecedent in relation to the excess by which the antecedent exceeds the consequent.

17. A ratio *ex aequali* arises when, there being several magnitudes and another set equal to them in multitude which taken two and two are in the same proportion, the first is to the last among the first magnitudes as the first is to the last among the second magnitudes. Or, in other words, it means taking the extreme terms by virtue of the removal of the intermediate terms.

18. A *perturbed proportion* arises when, there being three magnitudes and another set equal to them in multitude, antecedent is to consequent among the first magnitudes as antecedent is to consequent among the second magnitudes, while, the consequent is to a third among the first magnitudes as a third is to the antecedent among the second magnitudes.

Book V Propositions

Proposition 1

If any number of magnitudes are each the same multiple of the same number of other magnitudes, then the sum is that multiple of the sum.

Let any number of magnitudes AB and CD each be the same multiple of magnitudes E and F respectively.   V.Def.2

I say that the sum of AB and CD is the same multiple of the sum of E and F that AB is of E. Since AB is the same multiple of E that CD is of F, therefore there are as many magnitudes in AB equal to E as there are in CD equal to F.

Divide AB into magnitudes AG and GB equal to E, and divide CD into CH and HD equal to F. Then the number of the magnitudes AG and GB equals the number of the magnitudes CH and HD.

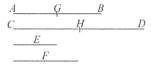

Now, since AG equals E, and CH equals F, therefore the sum of AG and CH equals the sum of E and F.

For the same reason GB equals E, and the sum of GB and HD equals the sum of E and F. Therefore, there are as many magnitudes in AB equal to E as there are in the sum of AB and CD equal to the sum of E and F.

Therefore, the sum of AB and CD is the same multiple of the sum of E and F that AB is of E.

Therefore, *if any number of magnitudes are each the same multiple of the same number of other magnitudes, then the sum is that multiple of the sum.*

Q.E.D.

Proposition 2

If a first magnitude is the same multiple of a second that a third is of a fourth, and a fifth also is the same multiple of the second that a sixth is of the fourth, then the sum of the first and fifth also is the same multiple of the second that the sum of the third and sixth is of the fourth.

Let a first magnitude AB be the same multiple of a second C that a third DE is of a fourth F, and let a fifth BG be the same multiple of the second C that a sixth EH is of the fourth F.

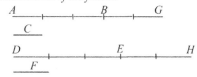

I say that the sum AG of the first and fifth is the same multiple of the second, C, that the sum DH of the third and sixth is of the fourth, F. Since AB is the same multiple of C that DE is of F,

therefore there are as many magnitudes in AB equal to C as there are in DE equal to F.

For the same reason there are as many magnitudes in BG equal to C as there are in EH equal to F. Therefore, there are as many magnitudes in the whole AG equal to C as there are in the whole DH equal to F.

Therefore, AG is the same multiple of C that DH is of F.

Therefore the sum AG of the first and fifth is the same multiple of the second, C, that the sum DH of the third and sixth is of the fourth, F.

Therefore, *if a first magnitude is the same multiple of a second that a third is of a fourth, and a fifth also is the same multiple of the second that a sixth is of the fourth, then the sum of the first and fifth also is the same multiple of the second that the sum of the third and sixth is of the fourth.*

Q.E.D.

Proposition 3

If a first magnitude is the same multiple of a second that a third is of a fourth, and if equimultiples are taken of the first and third, then the magnitudes taken also are equimultiples respectively, the one of the second and the other of the fourth.

Let a first magnitude A be the same multiple of a second B that a third C is of a fourth D, and let equimultiples EF and GH be taken of A and C.

V.Def.2

I say that EF is the same multiple of B that GH is of D.

Since EF is the same multiple of A that GH is of C, therefore there are as many magnitudes as in EF equal to A as there are in GH equal to C.

Divide EF into the magnitudes EK and KF equal to A, and divide GH into the magnitudes GL and LH equal to C. Then the number of the magnitudes EK and KF equals the number of the magnitudes GL and LH.

And, since A is the same multiple of B that C is of D, while EK equals A, and GL equals C, therefore EK is the same multiple of B that GL is of D.

For the same reason KF is the same multiple of B that LH is of D.

Since a first magnitude EK is the same multiple of a second B that a third GL is of a fourth D, and a fifth KF is the same multiple of the second B that a sixth LH is of the fourth D, therefore the sum EF of the first and fifth is the same multiple of the second B that the sum GH of the third and sixth is of the fourth D.

V.2

Therefore, *if a first magnitude is the same multiple of a second that a third is of a fourth, and if equimultiples are taken of the first and third, then the magnitudes taken also are equimultiples respectively, the one of the second and the other of the fourth.*

Q.E.D.

Proposition 4

If a first magnitude has to a second the same ratio as a third to a fourth, then any equimultiples whatever of the first and third also have the same ratio to any equimultiples whatever of the second and fourth respectively, taken in corresponding order.

Let a first magnitude A have to a second B the same ratio as a third C to a fourth D, and let equimultiples E and F be taken of A and C, and G and H other, arbitrary, equimultiples of B and D.
I say that E is to G as F is to H.

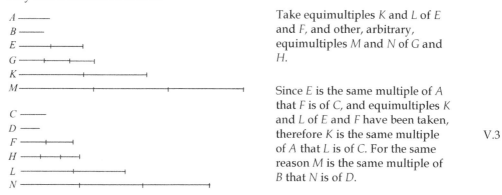

Take equimultiples K and L of E and F, and other, arbitrary, equimultiples M and N of G and H.

Since E is the same multiple of A that F is of C, and equimultiples K and L of E and F have been taken, therefore K is the same multiple of A that L is of C. For the same reason M is the same multiple of B that N is of D.   V.3

And, since A is to B as C is to D, and equimultiples K and L have been taken of A and C, and other, arbitrary, equimultiples M and N of B and D, therefore, if K is in excess of M, then L is in excess of N; if it is equal, equal; and if less, less.   V.Def.5

And K and L are equimultiples of E and F, and M and N are other, arbitrary, equimultiples of G and H, therefore E is to G as F is to H.   V.Def.5

Therefore, *if a first magnitude has to a second the same ratio as a third to a fourth, then any equimultiples whatever of the first and third also have the same ratio to any equimultiples whatever of the second and fourth respectively, taken in corresponding order.*

Q.E.D.

Proposition 5

If a magnitude is the same multiple of a magnitude that a subtracted part is of a subtracted part, then the remainder also is the same multiple of the remainder that the whole is of the whole.

Let the magnitude AB be the same multiple of the magnitude CD that the subtracted part AE is of the subtracted part CF.
I say that the remainder EB is also the same multiple of the remainder FD that the whole AB is of the whole CD.
Make CG so that EB is the same multiple of CG that AE is of CF.

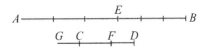

Then, since *AE* is the same multiple of *CF* that *EB* is of *GC*, therefore *AE* is the same multiple of *CF* that *AB* is of *GF*.

V.1

But, by the assumption, *AE* is the same multiple of *CF* that *AB* is of *CD*.
Therefore *AB* is the same multiple of each of the magnitudes *GF* and *CD*. Therefore *GF* equals *CD*.

Subtract *CF* from each. Then the remainder *GC* equals the remainder *FD*.
And, since *AE* is the same multiple of *CF* that *EB* is of *GC*, and *GC* equals *DF*, therefore *AE* is the same multiple of *CF* that *EB* is of *FD*.
But, by hypothesis, *AE* is the same multiple of *CF* that *AB* is of *CD*, therefore *EB* is the same multiple of *FD* that *AB* is of *CD*.
That is, the remainder *EB* is the same multiple of the remainder *FD* that the whole *AB* is of the whole *CD*.

Therefore, *If a magnitude is the same multiple of a magnitude that a subtracted part is of a subtracted part, then the remainder also is the same multiple of the remainder that the whole is of the whole.*

Q.E.D.

Proposition 6

If two magnitudes are equimultiples of two magnitudes, and any magnitudes subtracted from them are equimultiples of the same, then the remainders either equal the same or are equimultiples of them.

Let two magnitudes *AB* and *CD* be equimultiples of two magnitudes *E* and *F*, and let *AG* and *CH* subtracted from them be equimultiples of the same two *E* and *F*.
I say that the remainders *GB* and *HD* either equal *E* and *F* or are equimultiples of them.

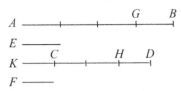

First, let *GB* equal *E*.
I say that *HD* also equals *F*.
Make *CK* equal to *F*.
Since *AG* is the same multiple of *E* that *CH* is of *F*, while *GB* equals *E*, and *KC* equals *F*, therefore *AB* is the same multiple of *E* that *KH* is of *F*.

V.2

But, by hypothesis, *AB* is the same multiple of *E* that *CD* is of *F*, therefore *KH* is the same multiple of *F* that *CD* is of *F*.
Since then each of the magnitudes *KH* and *CD* is the same multiple of *F*, therefore *KH* equals *CD*.
Subtract *CH* from each. Then the remainder *KC* equals the remainder *HD*.
But *F* equals *KC*, therefore *HD* also equals *F*.
Hence, if *GB* equals *E*, *HD* also equals *F*.
Similarly we can prove that, even if *GB* is a multiple of *E*, *HD* is also the same multiple of *F*.

Therefore, *if two magnitudes are equimultiples of two magnitudes, and any magnitudes subtracted from them are equimultiples of the same, then the remainders either equal the same or are equimultiples of them.*

Q.E.D.

Proposition 7

Equal magnitudes have to the same the same ratio; and the same has to equal magnitudes the same ratio.

Let A and B be equal magnitudes and C an arbitrary magnitude.
I say that each of the magnitudes A and B has the same ratio to C, and C has the same ratio to each of the magnitudes A and B.

Take equimultiples D and E of A and B, and take an arbitrary multiple F of C.
Then, since D is the same multiple of A that E is of B, and A equals B, therefore D equals E.

But F is another, arbitrary, magnitude. If therefore D is in excess of F, then E is also in excess of F; if equal, equal; and, if less, less.

And D and E are equimultiples of A and B, while F is another, arbitrary, multiple of C, therefore A is to C as B is to C.   V.Def.5

I say next that C also has the same ratio to each of the magnitudes A and B.
With the same construction, we can prove similarly that D equals E, and F is some other magnitude. If therefore F is in excess of D, it is also in excess of E; if equal, equal; and, if less, less.

And F is a multiple of C, while D and E are other, arbitrary, equimultiples of A and B, therefore C is to A as C is to B.   V.Def.5

Therefore, *equal magnitudes have to the same the same ratio; and the same has to equal magnitudes the same ratio.*

Q.E.D.

Corollary

From this it is manifest that, *if any magnitudes are proportional then they are proportional inversely.*

Proposition 8

Of unequal magnitudes, the greater has to the same a greater ratio than the less has; and the same has to the less a greater ratio than it has to the greater.

Let AB and C be unequal magnitudes, and let AB be greater, and let D be another, arbitrary, magnitude.

I say that AB has to D a greater ratio than C has to D, and D has to C a greater ratio than it has to AB.

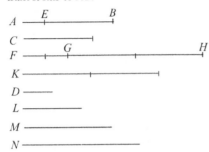

Since AB is greater than C, make EB equal to C. Then the less of the magnitudes AE and EB, if multiplied, will eventually be greater than D.   V.Def.4

First, let AE be less than EB. Let AE be multiplied, and let FG be a multiple of it which is greater than D. Make GH the same multiple of EB and K the same multiple of C that FG is of AE.

Take L double of D and M triple of it, and successive multiples increasing by one, until what is taken is the first multiple of D that is greater than K. Let it be taken, and   V.Def.4
let it be N which is quadruple of D and the first multiple of it greater than K.

Since K is less than N first, therefore K is not less than M.

And, since FG is the same multiple of AE that GH is of EB, therefore FG is the same   V.1
multiple of AE that FH is of AB.

But FG is the same multiple of AE that K is of C, therefore FH is the same multiple of AB that K is of C. Therefore FH and K are equimultiples of AB and C.

Again, since GH is the same multiple of EB that K is of C, and EB equals C, therefore GH equals K.

But K is not less than M, therefore neither is GH less than M.

And FG is greater than D, therefore the whole FH is greater than the sum of D and M.

But the sum of D and M equals N, inasmuch as M is triple D, and the sum of M and D is quadruple D, while N is also quadruple D, therefore the sum of M and D equals N.

But FH is greater than the sum of M and D, therefore FH is in excess of N, while K is not in excess of N.

And FH and K are equimultiples of AB and C, while N is another, arbitrary, multiple   V.Def.7
of D, therefore AB has to D a greater ratio than C has to D.

I say next, that D has to C a greater ratio than D has to AB.

With the same construction, we can prove similarly that N is in excess of K, while N is not in excess of FH.

And N is a multiple of D, while FH and K are other, arbitrary, equimultiples of AB V.Def.7
and C, therefore D has to C a greater ratio than D has to AB.

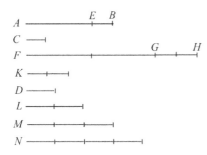

Next, let *AE* be greater than *EB*.

Then the less, *EB*, if multiplied, will eventually be greater than *D*. (V.Def.4)

Let it be multiplied, and let *GH* be a multiple of *EB* and greater than *D*. Make *FG* the same multiple of *AE*, and *K* the same multiple of *C* that *GH* is of *EB*.

Then we can prove similarly that *FH* and *K* are equimultiples of *AB* and *C*, and, similarly, take *N* the first multiple of *D* that is greater than *FG*, so that *FG* is again not less than *M*. (V.Def.4)

But *GH* is greater than *D*, therefore the whole *FH* is in excess of the sum of *D* and *M*, that is, of *N*.

Now *K* is not in excess of *N*, inasmuch as *FG* also, which is greater than *GH*, that is, than *K*, is not in excess of *N*.

And in the same manner, by following the above argument, we complete the demonstration.

Therefore, *of unequal magnitudes, the greater has to the same a greater ratio than the less has; and the same has to the less a greater ratio than it has to the greater.*

Q.E.D.

Proposition 9

Magnitudes which have the same ratio to the same equal one another; and magnitudes to which the same has the same ratio are equal.

Let each of the magnitudes *A* and *B* have the same ratio to *C*.
I say that *A* equals *B*.

Otherwise, each of the magnitudes *A* and *B* would not have the same ratio to *C*, but they do, therefore *A* equals *B*. V.8

Next, let *C* have the same ratio to each of the magnitudes *A* and *B*.
I say that *A* equals *B*.

Otherwise, *C* would not have the same ratio to each of the magnitudes *A* and *B*, but it does, therefore *A* equals *B*. V.8

Therefore, *magnitudes which have the same ratio to the same equal one another; and magnitudes to which the same has the same ratio are equal.*

Q.E.D.

Proposition 10

Of magnitudes which have a ratio to the same, that which has a greater ratio is greater; and that to which the same has a greater ratio is less.

Let *A* have to *C* a greater ratio than *B* has to *C*.
I say that *A* is greater than *B*.
If not, then *A* either equals *B* or is less than it.

Now *A* does not equal *B*, for in that case each of the magnitudes *A* and *B* would have the same ratio to *C*, but they do not, therefore *A* does not equal *B*.   V.7

Nor is *A* less than *B*, for in that case *A* would have to *C* a less ratio than *B* has to *C*, but it does not, therefore *A* is not less than *B*.   V.8

But it was proved not to be equal either, therefore *A* is greater than *B*.
Next, let *C* have to *B* a greater ratio than *C* has to *A*.
I say that *B* is less than *A*.
If not, it is either equal or greater.

Now *B* does not equal *A*, for in that case *C* would have the same ratio to each of the magnitudes *A* and *B*, but it does not, therefore *A* does not equal *B*.   V.7

Nor is *B* greater than *A*, for in that case *C* would have to *B* a less ratio than it has to *A*, but it does not, therefore *B* is not greater than *A*.   V.8

But it was proved that it is not equal either, therefore *B* is less than *A*.

Therefore, *of magnitudes which have a ratio to the same, that which has a greater ratio is greater; and that to which the same has a greater ratio is less.*

Q.E.D.

Proposition 11

Ratios which are the same with the same ratio are also the same with one another.

Let *A* be to *B* as *C* is to *D*, and let *C* be to *D* as *E* is to *F*.
I say that *A* is to *B* as *E* is to *F*.
Take equimultiples *G*, *H*, and *K* of *A*, *C*, and *E*, and take other, arbitrary, equimultiples *L*, *M*, and *N* of *B*, *D*, and *F*.

Then since *A* is to *B* as *C* is to *D*, and of *A* and *C* equimultiples *G* and *H* have been taken, and of *B* and *D* other, arbitrary, equimultiples *L* and *M*, therefore, if *G* is in excess of *L*, *H* is also in excess of *M*; if equal, equal; and if less, less.   V.Def.5

Again, since C is to D as E is to F, and of C and E equimultiples H and K have been taken, and of D and F other, arbitrary, equimultiples M and N, therefore, if H is in excess of M, K is also in excess of N; if equal, equal; and if less, less.

But we saw that, if H was in excess of M, G was also in excess of L; if equal, equal; and if less, less, so that, in addition, if G is in excess of L, K is also in excess of N; if equal, equal; and if less, less.   V.Def.5

And G and K are equimultiples of A and E, while L and N are other, arbitrary, equimultiples of B and F, therefore A is to B as E is to F.   V.Def.5

Therefore, *ratios which are the same with the same ratio are also the same with one another.*

Q.E.D.

Proposition 12

If any number of magnitudes are proportional, then one of the antecedents is to one of the consequents as the sum of the antecedents is to the sum of the consequents.

Let any number of magnitudes A, B, C, D, E, and F be proportional, so that A is to B as C is to D, and as E is to F.

I say that A is to B as the sum of A, C, and E is to the sum of B, D, and F.

```
A ——          C ——           E ——
B ——          D ——           F ——
G ————        H ————         K ————
L ——————      M ——————       N ——————
```

Take equimultiples G, H, and K of A, C, and E, and take other, arbitrary, equimultiples L, M, and N of B, D, and F.

Then since A is to B as C is to D, and as E is to F, and of A, C, and E equimultiples G, H, and K have been taken, and of B, D, and F other, arbitrary, equimultiples L, M, and N, therefore, if G is in excess of L, then H is also in excess of M, and K of N; if equal, equal; and if less, less. So that, in addition, if G is in excess of L, then the sum of G, H, and K is in excess of the sum of L, M, and N; if equal, equal; and if less, less.   V.Def.5

Now G and the sum of G, H, and K are equimultiples of A and the sum of A, C, and E, since, if any number of magnitudes are each the same multiple the same number of other magnitudes, then the sum is that multiple of the sum.   V.1

For the same reason L and the sum of L, M, and N are also equimultiples of B and the sum of B, D, and F, therefore A is to B as the sum of A, C, and E is to the sum of B, D, and F.   V.Def.5

Therefore, *if any number of magnitudes are proportional, then one of the antecedents is to one of the consequents as the sum of the antecedents is to the sum of the consequents.*

Q.E.D.

Proposition 13

If a first magnitude has to a second the same ratio as a third to a fourth, and the third has to the fourth a greater ratio than a fifth has to a sixth, then the first also has to the second a greater ratio than the fifth to the sixth.

Let a first magnitude *A* have to a second *B* the same ratio as a third *C* has to a fourth *D*, and let the third *C* have to the fourth *D* a greater ratio than a fifth *E* has to a sixth *F*.
I say that the first *A* also has to the second *B* a greater ratio than the fifth *E* to the sixth *F*.

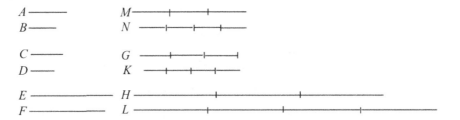

Since there are some equimultiples of *C* and *E*, and of *D* and *F* other equimultiples, such that the multiple of *C* is in excess of the multiple of *D*, while the multiple of *E* is not in excess of the multiple of *F*, let them be taken. Let *G* and *H* be equimultiples of *C* and *E*, and *K* and *L* other, arbitrary, equimultiples of *D* and *F*, so that *G* is in excess of *K*, but *H* is not in excess of *L*. Whatever multiple *G* is of *C*, let *M* also be that multiple of *A*, and, whatever multiple *K* is of *D*, let *N* also be that multiple of *B*.   V.Def.7

Now, since *A* is to *B* as *C* is to *D*, and of *A* and *C* equimultiples *M* and *G* have been taken, and of *B* and *D* other, arbitrary, equimultiples *N* and *K*, therefore, if *M* is in excess of *N*, *G* is also in excess of *K*; if equal, equal; and if less, less.   V.Def.5

But *G* is in excess of *K*, therefore *M* is also in excess of *N*.

But *H* is not in excess of *L*, and *M* and *H* are equimultiples of *A* and *E*, and *N* and *L* other, arbitrary, equimultiples of *B* and *F*, therefore *A* has to *B* a greater ratio than *E* has to *F*.   V.Def.7

Therefore, *if a first magnitude has to a second the same ratio as a third to a fourth, and the third has to the fourth a greater ratio than a fifth has to a sixth, then the first also has to the second a greater ratio than the fifth to the sixth.*

Q.E.D.

Proposition 14

If a first magnitude has to a second the same ratio as a third has to a fourth, and the first is greater than the third, then the second is also greater than the fourth; if equal, equal; and if less, less.

Let a first magnitude *A* have the same ratio to a second *B* as a third *C* has to a fourth *D*, and let *A* be greater than *C*.
I say that *B* is also greater than *D*.

Since *A* is greater than *C*, and *B* is another, arbitrary, magnitude, therefore *A* has to *B* a greater ratio than *C* has to *B*.	V.8
But *A* is to *B* as *C* is to *D*, therefore *C* has to *D* a greater ratio than *C* has to *B*.	V.13
But that to which the same has a greater ratio is less, therefore *D* is less than *B*, so that *B* is greater than *D*.	V.10

Similarly we can prove that, if *A* equals *C*, then *B* equals *D*, and, if *A* is less than *C*, then *B* is less than *D*.

Therefore, *if a first magnitude has to a second the same ratio as a third has to a fourth, and the first is greater than the third, then the second is also greater than the fourth; if equal, equal; and if less, less.*

Q.E.D.

Proposition 15

Parts have the same ratio as their equimultiples.
Let *AB* be the same multiple of *C* that *DE* is of *F*.
I say that *C* is to *F* as *AB* is to *DE*.
Since *AB* is the same multiple of *C* that *DE* is of *F*, as many magnitudes as there are in *AB* equal to *C*, there are also in *DE* equal to *F*.

Divide *AB* into the magnitudes *AG*, *GH*, and *HB* equal to *C*, and divide *DE* into the magnitudes *DK*, *KL*, and *LE* equal to *F*.

Then the number of the magnitudes *AG*, *GH*, and *HB* equals the number of the magnitudes *DK*, *KL*, and *LE*.

And, since *AG*, *GH*, and *HB* equal one another, and *DK*, *KL*, and *LE* also equal one another, therefore *AG* is to *DK* as *GH* is to *KL*, and as *HB* is to *LE*.	V.7
Therefore one of the antecedents is to one of the consequents as the sum of the antecedents is to the sum of the consequents. Therefore *AG* is to *DK* as *AB* is to *DE*.	V.12

But *AG* equals *C* and *DK* equals *F*, therefore *C* is to *F* as *AB* is to *DE*.
Therefore, *parts have the same ratio as their equimultiples.*

Q.E.D.

Proposition 16

If four magnitudes are proportional, then they are also proportional alternately.
Let *A*, *B*, *C*, and *D* be four proportional magnitudes, so that *A* is to *B* as *C* is to *D*.
I say that they are also so alternately, that is *A* is to *C* as *B* is to *D*.   V.Def.12

Take equimultiples *E* and *F* of *A* and *B*, and take other, arbitrary, equimultiples *G* and *H* of *C* and *D*.

Then, since *E* is the same multiple of *A* that *F* is of *B*, and parts have the same ratio as their equimultiples, therefore *A* is to *B* as *E* is to *F*.   V.15

But *A* is to *B* as *C* is to *D*, therefore *C* is to *D* also as *E* is to *F*.   V.11

Again, since *G* and *H* are equimultiples of *C* and *D*, therefore *C* is to *D* as *G* is to *H*.   V.15

But *C* is to *D* as *E* is to *F*, therefore as *E* is to *F* also as *G* is to *H*.   V.11

But, if four magnitudes are proportional, and the first is greater than the third, then the second is also greater than the fourth; if equal, equal; and if less, less.   V.14

Therefore, if *E* is in excess of *G*, *F* is also in excess of *H*; if equal, equal; and if less, less.

Now *E* and *F* are equimultiples of *A* and *B*, and *G* and *H* other, arbitrary, equimultiples of *C* and *D*, therefore *A* is to *C* as *B* is to *D*.   V.Def.5

Therefore, *if four magnitudes are proportional, then they are also proportional alternately.*

Q.E.D.

Proposition 17

If magnitudes are proportional taken jointly, then they are also proportional taken separately.

Let *AB*, *BE*, *CD*, and *DF* be magnitudes proportional taken jointly, so that *AB* is to *BE* as *CD* is to *DF*.   V.Def.14

I say that they are also proportional taken separately, that is, *AE* is to *EB* as *CF* is to *DF*.   V.Def.15

Take equimultiples *GH*, *HK*, *LM*, and *MN* of *AE*, *EB*, *CF*, and *FD*, and take other, arbitrary, equimultiples, *KO* and *NP* of *EB* and *FD*.

Then, since *GH* is the same multiple of *AE* that *HK* is of *EB*, therefore *GH* is the same multiple of *AE* that *GK* is of *AB*.   V.1

But *GH* is the same multiple of *AE* that *LM* is of *CF*, therefore *GK* is the same multiple of *AB* that *LM* is of *CF*.

Again, since *LM* is the same multiple of *CF* that *MN* is of *FD*, therefore *LM* is the same multiple of *CF* that *LN* is of *CD*.   V.1

But *LM* was the same multiple of *CF* that *GK* is of *AB*, therefore *GK* is the same multiple of *AB* that *LN* is of *CD*.

Therefore *GK* and *LN* are equimultiples of *AB* and *CD*.

Again, since *HK* is the same multiple of *EB* that *MN* is of *FD*, and *KO* is also the same multiple of *EB* that *NP* is of *FD*, therefore the sum *HO* is also the same multiple of *EB* that *MP* is of *FD*.

V.2

And, since *AB* is to *BE* as *CD* is to *DF*, and of *AB* and *CD* equimultiples *GK* and *LN* have been taken, and of *EB* and *FD* equimultiples *HO* and *MP*, therefore, if *GK* is in excess of *HO*, and *LN* is also in excess of *MP*; if equal, equal; and if less, less.

Let *GK* be in excess of *HO*. Subtract *HK* from each. Therefore *GH* is also in excess of *KO*.

But we saw that, if *GK* was in excess of *HO*, then *LN* was also in excess of *MP*, therefore *LN* is also in excess of *MP*, and, if *MN* is subtracted from each, then *LM* is also in excess of *NP*, so that, if *GH* is in excess of *KO*, then *LM* is also in excess of *NP*. Similarly we can prove that, if *GH* equals *KO*, then *LM* also equals *NP*; and if less, less.

And *GH* and *LM* are equimultiples of *AE* and *CF*, while *KO* and *NP* are other, arbitrary, equimultiples of *EB* and *FD*, therefore *AE* is to *EB* as *CF* is to *FD*.

V.Def.5

Therefore, *if magnitudes are proportional taken jointly, then they are also proportional taken separately.*

Q.E.D.

Proposition 18

If magnitudes are proportional taken separately, then they are also proportional taken jointly.

Let *AE*, *EB*, *CF*, and *FD* be magnitudes proportional taken separately, so that *AE* is to *EB* as *CF* is to *FD*.

V.Def.15

I say that they are also proportional taken jointly, that is, *AB* is to *BE* as *CD* is to *FD*.

V.Def.14

For, if *CD* is not to *DF* as *AB* is to *BE*, then *AB* is to *BE* as *CD* is either to some magnitude less than *DF* or to a greater.

First, let it be in that ratio to a less magnitude *DG*.

Then, since *AB* is to *BE* as *CD* is to *DG*, they are magnitudes proportional taken jointly, so that they are also proportional taken separately. Therefore *AE* is to *EB* as *CG* is to *GD*.

V.17

But also, by hypothesis, *AE* is to *EB* as *CF* is to *FD*. Therefore *CG* is to *GD* as *CF* is to *FD*.

V.11

But the first *CG* is greater than the third *CF*, therefore the second *GD* is also greater than the fourth *FD*.

V.14

But it is also less, which is impossible. Therefore *AB* is to *BE* as *CD* is not to a less magnitude than *FD*.

Similarly we can prove that neither is it in that ratio to a greater, it is therefore in that ratio to *FD* itself.

Therefore, *if magnitudes are proportional taken separately, then they are also proportional taken jointly.*

Q.E.D.

Proposition 19

If a whole is to a whole as a part subtracted is to a part subtracted, then the remainder is also to the remainder as the whole is to the whole.

Let the whole *AB* be to the whole *CD* as the part *AE* subtracted is to the part *CF* subtracted.

I say that the remainder *EB* is also to the remainder *FD* as the whole *AB* is to the whole *CD*.

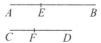

Since *AB* is to *CD* as *AE* is to *CF*, therefore alternately, *BA* is to *AE* as *DC* is to *CF*.   V.16

And, since the magnitudes are proportional taken jointly, they are also proportional taken separately, that is, *BE* is to *EA* as *DF* is to *CF*, and, alternately, *BE* is to *DF* as *EA* is to *FC*.   V.17   V.16

But, by hypothesis, *AE* is to *CF* as is the whole *AB* to the whole *CD*.

Therefore the remainder *EB* is also to the remainder *FD* as the whole *AB* is to the whole *CD*.   V.11

Therefore *if a whole is to a whole as a part subtracted is to a part subtracted, then the remainder is also to the remainder as the whole is to the whole.*

Q.E.D.

Corollary

From this it is manifest that, *if magnitudes are proportional taken jointly, then they are also proportional in conversion.*   V.Def.16

Proposition 20

If there are three magnitudes, and others equal to them in multitude, which taken two and two are in the same ratio, and if ex aequali the first is greater than the third, then the fourth is also greater than the sixth; if equal, equal, and; if less, less.

Let there be three magnitudes *A*, *B*, and *C*, and others *D*, *E*, and *F* equal to them in multitude, which taken two and two are in the same ratio, so that *A* is to *B* as *D* is to *E*, and *B* is to *C* as *E* is to *F*.

Let *A* be greater than *C ex aequali*.

I say that *D* is also greater than *F*; if *A* equals *C*, equal; and, if less, less.

```
A ———————   D ———
B ———        E ——
C ————       F ——
```

Since *A* is greater than *C*, and *B* is some other magnitude, and the greater has to the same a greater ratio than the less has, therefore *A* has to *B* a greater ratio than *C* has to *B*. V.8

But *A* is to *B* as *D* is to *E*, and, *C* is to *B*, inversely, as *F* is to *E*, therefore *D* has to *E* a greater ratio than *F* has to *E*. V.7.Cor V.13

But, of magnitudes which have a ratio to the same, that which has a greater ratio is greater, therefore *D* is greater than *F*. V.10

Similarly we can prove that, if *A* equals *C*, then *D* also equals *F*, and if less, less.

Therefore, *if there are three magnitudes, and others equal to them in multitude, which taken two and two are in the same ratio, and if ex aequali the first is greater than the third, then the fourth is also greater than the sixth; if equal, equal, and; if less, less.*

Q.E.D.

Proposition 21

If there are three magnitudes, and others equal to them in multitude, which taken two and two together are in the same ratio, and the proportion of them is perturbed, then, if ex aequali the first magnitude is greater than the third, then the fourth is also greater than the sixth; if equal, equal; and if less, less.

Let there be three magnitudes *A*, *B*, and *C*, and others *D*, *E*, and *F* equal to them in multitude, which taken two and two are in the same ratio, and let the proportion of them be perturbed, so that *A* is to *B* as *E* is to *F*, and *B* is to *C* as *D* is to *E*. V.Def.18

Let *A* be greater than *C* ex aequali.
I say that *D* is also greater than *F*; if *A* equals *C*, equal; and if less, less.

```
A ——————      D ————————
B ———————     E ———————————
C ————————    F ——————————
```

Since *A* is greater than *C*, and *B* is some other magnitude, therefore *A* has to *B* a greater ratio than *C* has to *B*. V.8

But *A* is to *B* as *E* is to *F*, and, inversely, *C* is to *B* as *E* is to *D*. Therefore also *E* has to *F* a greater ratio than *E* has to *D*. V.7.Cor V.13

But that to which the same has a greater ratio is less, therefore *F* is less than *D*, therefore *D* is greater than *F*. V.10

Similarly we can prove that, if *A* equals *C*, then *D* also equals *F*; and if less, less.

Therefore, *if there are three magnitudes, and others equal to them in multitude, which taken two and two together are in the same ratio, and the proportion of them is perturbed, then, if ex aequali the first magnitude is greater than the third, then the fourth is also greater than the sixth; if equal, equal; and if less, less.*

Q.E.D.

Proposition 22

If there are any number of magnitudes whatever, and others equal to them in multitude, which taken two and two together are in the same ratio, then they are also in the same ratio ex aequali.

Let there be any number of magnitudes A, B, and C, and others D, E, and F equal to them in multitude, which taken two and two together are in the same ratio, so that A is to B as D is to E, and B is to C as E is to F.

I say that they are also in the same ratio *ex aequali*, that is, A is to C as D is to F.   V.Def.17

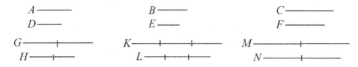

Take equimultiples G and H of A and D, and take other, arbitrary, equimultiples K and L of B and E, and, further, take other, arbitrary, equimultiples M and N of C and F.

Then, since A is to B as D is to E, and of A and D equimultiples G and H have been taken, and of B and E other, arbitrary, equimultiples K and L, therefore G is to K as H is to L.   V.4

For the same reason also K is to M as L is to N.

Since, then, there are three magnitudes G, K, and M, and others H, L, and N equal to them in multitude, which taken two and two together are in the same ratio, therefore, *ex aequali*, if G is in excess of M, H is also in excess of N; if equal, equal; and if less, less.   V.20

And G and H are equimultiples of A and D, and M and N other, arbitrary, equimultiples of C and F.

Therefore A is to C as D is to F.   V.Def.5

Therefore, *if there are any number of magnitudes whatever, and others equal to them in multitude, which taken two and two together are in the same ratio, then they are also in the same ratio ex aequali.*

Q.E.D.

Book VI

Definitions

1. *Similar* rectilinear figures are such as have their angles severally equal and the sides about the equal angles proportional.

2. Two figures are *reciprocally related* when the sides about corresponding angles are reciprocally proportional.

3. A straight line is said to have been *cut in extreme and mean ratio* when, as the whole line is to the greater segment, so is the greater to the less.

4. The *height* of any figure is the perpendicular drawn from the vertex to the base.

Book VI Propositions

Proposition 1

Triangles and parallelograms which are under the same height are to one another as their bases.

Let ABC, ACD be triangles and EC, EF parallelograms under the same height;
I say that the base BC is to the base CD as the triangle ACB is to the triangle ACD, and as the parallelogram CE is to the parallelogram CF.

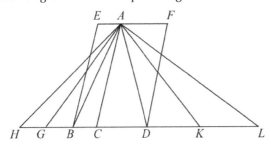

Produce BD in both directions to the points H and L. Make any number of straight lines BG and GH equal to the base CB, and any number of straight lines DK and KL equal to the base CD. Join AG, AH, AK, and AL.

I.3

Then, since CB, BG, and GH equal one another, the triangles ACB, ABG, and AGH also equal one another.

I.38

Therefore, whatever multiple the base CH is of the base CB, the triangle ACH is also that multiple of the triangle ACB.

For the same reason, whatever multiple the base CL is of the base CD, the triangle ACL is also that multiple of the triangle ACD. And, if the base CH equals the base CL, then the triangle ACH also equals the triangle ACL; if the base CH is in excess of the base CL, the triangle ACH is also in excess of the triangle ACL; and, if less, less.

I.38

Thus, there being four magnitudes, namely two bases CB and CD, and two triangles ACB and ACD, equimultiples have been taken of the base CB and the triangle ACB, namely the base CH and the triangle ACH, and other, arbitrary, equimultiples of the base CD and the triangle ADC, namely the base CL and the triangle ACL, and it has been proved that, if the base CH is in excess of the base CL, the triangle ACH is also in excess of the triangle ACL; if equal, equal; and, if less, less. Therefore the base CB is to the base CD as the triangle ACB is to the triangle ACD.

V.Def.5

Next, since the parallelogram CE is double the triangle ACB, and the parallelogram FC is double the triangle ACD, and parts have the same ratio as their equimultiples, therefore the triangle ACB is to the triangle ACD as the parallelogram CE is to the parallelogram FC.

I.41
V.15

Since, then, it was proved that the base CB is to CD as the triangle ACB is to the triangle ACD, and the triangle ACB is to the triangle ACD as the parallelogram CE is to the parallelogram CF, therefore also the base CB is to the base CD as the parallelogram CE is to the parallelogram FC. V.11

Therefore, triangles and parallelograms which are under the same height are to one another as their bases.

Q.E.D.

Proposition 2

If a straight line is drawn parallel to one of the sides of a triangle, then it cuts the sides of the triangle proportionally; and, if the sides of the triangle are cut proportionally, then the line joining the points of section is parallel to the remaining side of the triangle.

Let DE be drawn parallel to BC, one of the sides of the triangle ABC.
I say that BD is to AD as CE is to AE.

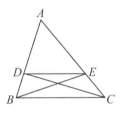

Join BE and CD.

Therefore the triangle BDE equals the triangle CDE, for they are on the same base DE and in the same parallels DE and BC. I.37

And ADE is another triangle.

But equals have the same ratio to the same, therefore the triangle BDE is to the triangle ADE as the triangle CDE is to the triangle ADE. V.7

But the triangle BDE is to ADE as BD is to AD, for, being under the same height, the perpendicular drawn from E to AB, they are to one another as their bases. VI.1

For the same reason, the triangle CDE is to ADE as CE is to AE.

Therefore BD is to AD also as CE is to AE. V.11

Next, let the sides AB and AC of the triangle ABC be cut proportionally, so that BD is to AD as CE is to AE. Join DE.
I say that DE is parallel to BC.

With the same construction, since BD is to AD as CE is to AE, but BD is to AD as the triangle BDE is to the triangle ADE, and CE is to AE as the triangle CDE is to the triangle ADE, therefore the triangle BDE is to the triangle ADE as the triangle CDE is to the triangle ADE. VI.1 V.11

Therefore each of the triangles BDE and CDE has the same ratio to ADE.

Therefore the triangle BDE equals the triangle CDE, and they are on the same base DE. V.9

But equal triangles which are on the same base are also in the same parallels. I.39

Therefore DE is parallel to BC.

Therefore, if a straight line is drawn parallel to one of the sides of a triangle, then it cuts the sides of the triangle proportionally; and, if the sides of the triangle are cut proportionally, then the line joining the points of section is parallel to the remaining side of the triangle.

Q.E.D.

Proposition 3

If an angle of a triangle is bisected by a straight line cutting the base, then the segments of the base have the same ratio as the remaining sides of the triangle; and, if segments of the base have the same ratio as the remaining sides of the triangle, then the straight line joining the vertex to the point of section bisects the angle of the triangle.

Let ABC be a triangle, and let the angle BAC be bisected by the straight line AD.
I say that DB is to DC as AB is to AC.

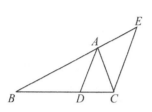

Draw CE through C parallel to DA, and carry AB through to meet it at E.	I.31
Then, since the straight line AC falls upon the parallels AD and EC, the angle ACE equals the angle CAD.	I.29
But the angle CAD equals the angle BAD by hypothesis, therefore the angle BAD also equals the angle ACE.	
Again, since the straight line BAE falls upon the parallels AD and EC, the exterior angle BAD equals the interior angle AEC.	I.29
But the angle ACE was also proved equal to the angle BAD, therefore the angle ACE also equals the angle AEC, so that the side AE also equals the side AC.	I.6
And, since AD is parallel to EC, one of the sides of the triangle BCE, therefore, proportionally DB is to DC as AB is to AE.	VI.2
But AE equals AC, therefore DB is to DC as AB is to AC.	V.7

Next, let DB be to DC as AB is to AC. Join AD.
I say that the straight line AD bisects the angle BAC.

With the same construction, since DB is to DC as AB is to AC, and also DB is to DC as AB is to AE, for AD is parallel to EC, one of the sides of the triangle BCE, therefore also AB is to AC as AB is to AE.	VI.2 V.11
Therefore AC equals AE, so that the angle AEC also equals the angle ACE.	V.9, I.5
But the angle AEC equals the exterior angle BAD, and the angle ACE equals the alternate angle CAD, therefore the angle BAD also equals the angle CAD.	I.29

Therefore the straight line AD bisects the angle BAC.

Therefore, *if an angle of a triangle is bisected by a straight line cutting the base, then the segments of the base have the same ratio as the remaining sides of the triangle; and, if segments of the base have the same ratio as the remaining sides of the triangle, then the straight line joining the vertex to the point of section bisects the angle of the triangle.*

Q.E.D.

Proposition 4

In equiangular triangles the sides about the equal angles are proportional where the corresponding sides are opposite the equal angles.

Let ABC and DCE be equiangular triangles having the angle ABC equal to the angle DCE, the angle BAC equal to the angle CDE, and the angle ACB equal to the angle CED.

I say that in the triangles ABC and DEC the sides about the equal angles are proportional where the corresponding sides are opposite the equal angles.

Let BC be placed in a straight line with CE.

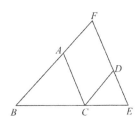

Then, since the sum of the angles ABC and ACB is less than two right angles, and the angle ACB equals the angle DEC, therefore the sum of the angles ABC and DEC is less than two right angles. Therefore BA and ED, when produced, will meet. Let them be produced and meet at F.	I.17 I.Post.5
Now, since the angle DCE equals the angle ABC, DC is parallel to FB. Again, since the angle ACB equals the angle DEC, AC is parallel to FE.	I.28
Therefore FACD is a parallelogram, therefore FA equals DC, and AC equals FD.	I.34
And, since AC is parallel to a side FE of the triangle FBE, therefore BA is to AF as BC is to CE.	VI.2
But AF equals CD, therefore BA is to CD as BC is to CE, and alternately AB is to BC as DC is to CE.	V.7 V.16
Again, since CD is parallel to BF, therefore BC is to CE as FD is to DE.	VI.2
But FD equals AC, therefore BC is to CE as AC is to DE, and alternately BC is to CA as CE is to ED.	V.7 V.16
Since then it was proved that AB is to BC as DC is to CE, and BC is to CA as CE is to ED, therefore, *ex aequali*, BA is to AC as CD is to DE.	V.22

Therefore, *in equiangular triangles the sides about the equal angles are proportional where the corresponding sides are opposite the equal angles.*

Q.E.D.

Proposition 5

If two triangles have their sides proportional, then the triangles are equiangular with the equal angles opposite the corresponding sides.

Let ABC and DEF be two triangles having their sides proportional, so that AB is to BC as DE is to EF, BC is to CA as EF is to FD, and further BA is to AC as ED is to DF.

I say that the triangle ABC is equiangular with the triangle DEF where the equal angles are opposite the corresponding sides, namely the angle ABC equals the angle DEF, the angle BCA equals the angle EFD, and the angle BAC equals the angle EDF.

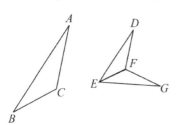

Construct the angle FEG equal to the angle CBA and the angle EFG equal to the angle BCA on the straight line EF and at the points E and F on it. Therefore the remaining angle at A equals the remaining angle at G.   I.23, I.32

Therefore the triangle ABC is equiangular with the triangle GEF.

Therefore in the triangles ABC and GEF the sides about the equal angles are proportional where the corresponding sides are opposite the equal angles, therefore AB is to BC as GE is to EF.   VI.4

But, by hypothesis, AB is to BC as DE to EF, therefore DE is to EF as GE is to EF.   V.11

Therefore each of the straight lines DE and GE has the same ratio to EF, therefore DE equals GE.   V.9

For the same reason DF also equals GF.

Then since DE equals GE, and EF is common, the two sides DE and EF equal the two sides GE and EF, and the base DF equals the base GF, therefore the angle DEF equals the angle GEF, and the triangle DEF equals the triangle GEF, and the remaining angles equal the remaining angles, namely those opposite the equal sides.   I.8 I.4

Therefore the angle DFE also equals the angle GFE, and the angle EDF equals the angle EGF.

And, since the angle DEF equals the angle GEF, and the angle GEF equals the angle ABC, therefore the angle ABC also equals the angle DEF.

For the same reason the angle ACB also equals the angle DFE, and further, the angle at A equals the angle at D, therefore the triangle ABC is equiangular with the triangle DEF.

Therefore, *if two triangles have their sides proportional, then the triangles are equiangular with the equal angles opposite the corresponding sides.*

    Q.E.D.

Proposition 6

If two triangles have one angle equal to one angle and the sides about the equal angles proportional, then the triangles are equiangular and have those angles equal opposite the corresponding sides.

Let *ABC* and *DEF* be two triangles having one angle *BAC* equal to one angle *EDF* and the sides about the equal angles proportional, so that BA is to AC as ED is to DF.
I say that the triangle *ABC* is equiangular with the triangle *DEF*, and has the angle *ABC* equal to the angle *DEF*, and the angle *ACB* equal to the angle *DFE*.

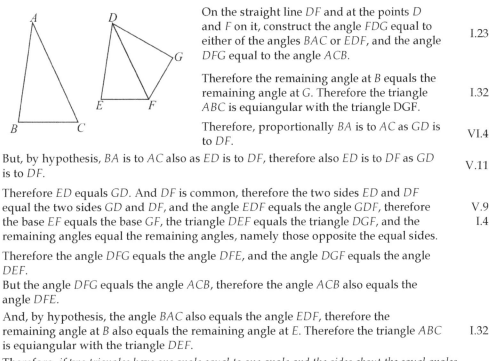

On the straight line *DF* and at the points *D* and *F* on it, construct the angle *FDG* equal to either of the angles *BAC* or *EDF*, and the angle *DFG* equal to the angle *ACB*.	I.23
Therefore the remaining angle at *B* equals the remaining angle at *G*. Therefore the triangle *ABC* is equiangular with the triangle *DGF*.	I.32
Therefore, proportionally *BA* is to *AC* as *GD* is to *DF*.	VI.4

But, by hypothesis, *BA* is to *AC* also as *ED* is to *DF*, therefore also *ED* is to *DF* as *GD* is to *DF*. V.11

Therefore *ED* equals *GD*. And *DF* is common, therefore the two sides *ED* and *DF* equal the two sides *GD* and *DF*, and the angle *EDF* equals the angle *GDF*, therefore the base *EF* equals the base *GF*, the triangle *DEF* equals the triangle *DGF*, and the remaining angles equal the remaining angles, namely those opposite the equal sides. V.9 I.4

Therefore the angle *DFG* equals the angle *DFE*, and the angle *DGF* equals the angle *DEF*.

But the angle *DFG* equals the angle *ACB*, therefore the angle *ACB* also equals the angle *DFE*.

And, by hypothesis, the angle *BAC* also equals the angle *EDF*, therefore the remaining angle at *B* also equals the remaining angle at *E*. Therefore the triangle *ABC* is equiangular with the triangle *DEF*. I.32

Therefore, *if two triangles have one angle equal to one angle and the sides about the equal angles proportional, then the triangles are equiangular and have those angles equal opposite the corresponding sides.*

Q.E.D.

Proposition 7

If two triangles have one angle equal to one angle, the sides about other angles proportional, and the remaining angles either both less or both not less than a right angle, then the triangles are equiangular and have those angles equal the sides about which are proportional.

Let *ABC* and *DEF* be two triangles having one angle equal to one angle, the angle *BAC* equal to the angle *EDF*, the sides about other angles *ABC* and *DEF* proportional, so that *AB* is to *BC* as *DE* is to *EF*. And, first, each of the remaining angles at *C* and *F* less than a right angle. I say that the triangle *ABC* is equiangular with the triangle *DEF*, the angle *ABC* equals the angle *DEF*, and the remaining angle, namely the angle at *C*, equals the remaining angle, the angle at *F*.

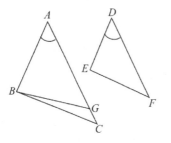

If the angle *ABC* does not equal the angle *DEF*, then one of them is greater.

Let the angle *ABC* be greater. Construct the angle *ABG* equal to the angle *DEF* on the straight line *AB* and at the point *B* on it. | I.23

Then, since the angle *A* equals *D*, and the angle *ABG* equals the angle *DEF*, therefore the remaining angle *AGB* equals the remaining angle *DFE*. | I.32

Therefore the triangle *ABG* is equiangular with the triangle *DEF*.

Therefore *AB* is to *BG* as *DE* is to *EF*. | VI.4

But, by hypothesis, *DE* is to *EF* as *AB* is to *BC*, therefore *AB* has the same ratio to each of the straight lines *BC* and *BG*. Therefore *BC* equals *BG*, so that the angle at *C* also equals the angle *BGC*. | V.11, V.9, I.5

But, by hypothesis, the angle at *C* is less than a right angle, therefore the angle *BGC* is also less than a right angle, so that the angle *AGB* adjacent to it is greater than a right angle. | I.13

And it was proved equal to the angle at *F*, therefore the angle at *F* is also greater than a right angle. But it is by hypothesis less than a right angle, which is absurd.
Therefore the angle *ABC* is not unequal to the angle *DEF*. Therefore it equals it.

But the angle at *A* also equals the angle at *D*, therefore the remaining angle at *C* equals the remaining angle at *F*. | I.32

Therefore the triangle *ABC* is equiangular with the triangle *DEF*.

Next let each of the angles *C* and *F* be supposed not less than a right angle. I say again that, in this case too, the triangle *ABC* is equiangular with the triangle *DEF*.

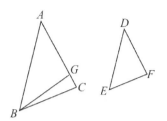

With the same construction, we can prove similarly that BC equals BG, so that the angle at C also equals the angle BGC. act   I.5

But the angle at C is not less than a right angle, therefore neither is the angle BGC less than a right angle.

Thus in the triangle BGC the sum of two angles is not less than two right angles, which is impossible.   I.17

Therefore, once more, the angle ABC is not unequal to the angle DEF. Therefore it equals it.

But the angle at A also equals the angle at D, therefore the remaining angle at C equals the remaining angle at F.   I.32

Therefore the triangle ABC is equiangular with the triangle DEF.

Therefore, *if two triangles have one angle equal to one angle, the sides about other angles proportional, and the remaining angles either both less or both not less than a right angle, then the triangles are equiangular and have those angles equal the sides about which are proportional.*

Q.E.D.

Proposition 8

If in a right-angled triangle a perpendicular is drawn from the right angle to the base, then the triangles adjoining the perpendicular are similar both to the whole and to one another.

Let ABC be a right-angled triangle having the angle BAC right, and let AD be drawn from A perpendicular to BC. I say that each of the triangles DBA and DAC is similar to the whole ABC, and, further, they are similar to one another.

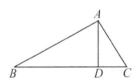

Since the angle BAC equals the angle BDA, for each is right, and the angle at B is common to the two triangles ABC and DBA, therefore the remaining angle ACB equals the remaining angle DAB. Therefore the triangle ABC is equiangular with the triangle DBA.   I.32

Therefore BC, which is opposite the right angle in the triangle ABC, is to BA, which is opposite the right angle in the triangle DBA, as AB, which is opposite the angle at C in the triangle ABC, is to DB, which is opposite the equal angle BAD in the triangle DBA, and also as AC is to DA, which is opposite the angle at B common to the two triangles.   VI.4

Therefore the triangle ABC is both equiangular to the triangle DBA and has the sides about the equal angles proportional.

Therefore the triangle ABC is similar to the triangle DBA.   VI.Def.1

In the same manner we can prove that the triangle DAC is also similar to the triangle ABC. Therefore each of the triangles DBA and DAC is similar to the whole ABC.

I say next that the triangles DBA and DAC are also similar to one another.

Since the right angle *BDA* equals the right angle *ADC*, and moreover the angle *DAB* was also proved equal to the angle at *C*, therefore the remaining angle at *B* also equals the remaining angle *DAC*. Therefore the triangle *DBA* is equiangular with the triangle *ADC*.

I.32

Therefore *BD*, which is opposite the angle *DAB* in the triangle *DBA*, is to *AD*, which is opposite the angle at *C* in the triangle *DAC* equal to the angle *DAB*, as *AD*, itself which is opposite the angle at *B* in the triangle *DBA*, is to *CD*, which is opposite the angle *DAC* in the triangle *DAC* equal to the angle at *B*, and also as *BA* is to *AC*, these sides opposite the right angles. Therefore the triangle *DBA* is similar to the triangle *DAC*.

VI.4
VI.Def.1

Therefore, *if in a right-angled triangle a perpendicular is drawn from the right angle to the base, then the triangles adjoining the perpendicular are similar both to the whole and to one another.*

Q.E.D.

Corollary
From this it is clear that, if in a right-angled triangle a perpendicular is drawn from the right angle to the base, then the straight line so drawn is a mean proportional between the segments of the base.

Proposition 9

To cut off a prescribed part from a given straight line.

Let *AB* be the given straight line.
It is required to cut off from *AB* a prescribed part.
Let the third part be that prescribed.

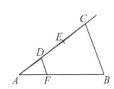

Draw a straight line *AC* through from A containing with *AB* any angle. Take a point *D* random on *AC*, and make *DE* and *EC* equal to *AD*.

I.3

Join *CB*, and draw *DF* through *D* parallel to it.

I.31

Then, since *DF* is parallel to a side *CB* of the triangle *ABC*, therefore, proportionally, *AD* is to *DC* as *AF* is to *FB*.

VI.2

But *DC* is double *AD*, therefore *FB* is also double *AF*, therefore *AB* is triple of *AF*. Therefore from the given straight line *AB* the prescribed third part *AF* has been cut off.

Q.E.F.

Proposition 10

To cut a given uncut straight line similarly to a given cut straight line.

Let *AB* be the given uncut straight line, and *AC* the straight line cut at the points *D* and *E*, and let them be so placed as to contain any angle. Join *CB*, and draw *DF* and *EG* through *D* and *E* parallel to *CB*, and draw *DHK* through *D* parallel to *AB*.

I.31

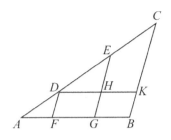

Therefore each of the figures *FH* and *HB* is a parallelogram. Therefore *DH* equals *FG* and *HK* equals *GB*. I.34

Now, since the straight line *EH* is parallel to a side *CK* of the triangle *DCK*, therefore, proportionally, *DE* is to *EC* as *DH* is to *HK*. VI.2

But *DH* equals *FG*, and *HK* equals *GB*, therefore *DE* is to *EC* as *FG* is to *GB*. V.7

Again, since *DF* is parallel to a side *EG* of the triangle *AEG*, therefore, proportionally, *AD* is to *DE* as *AF* is to *FG*. VI.2

But it was also proved that *DE* is to *EC* as *FG* is to *GB*, therefore *DE* is to *EC* as *FG* is to *GB*, and *AD* is to *DE* as *AF* is to *FG*.

Therefore the given uncut straight line *AB* has been cut similarly to the given cut straight line *AC*.

Q.E.F.

Proposition 11

To find a third proportional to two given straight lines.

Let *AB* and *AC* be the two given straight lines, and let them be placed so as to contain any angle.

It is required to find a third proportional to *AB* and *AC*.

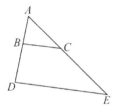

Produce them to the points *D* and *E*, and make *BD* equal to *AC*. Join *BC*, and draw *DE* through *D* parallel to it. I.3
I.31

Then since *BC* is parallel to a side *DE* of the triangle *ADE*, therefore, proportionally, *AB* is to *BD* as *AC* is to *CE*. VI.2

But *BD* equals *AC*, therefore *AB* is to *AC* as *AC* is to *CE*. V.7

Therefore a third proportional *CE* has been found to two given straight lines *AB* and *AC*.

Q.E.F.

Proposition 12

To find a fourth proportional to three given straight lines.

Let A and B and C be the three given straight lines.
It is required to find a fourth proportional to A, B, and C.

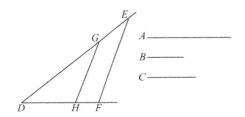

Set out two straight lines DE and DF containing any angle EDF. Make DG equal to A, GE equal to B, and DH equal to C. Join GH, and draw EF through E parallel to it.

I.3
I.31

Then since GH is parallel to a side EF of the triangle DEF, therefore DG is to GE as DH is to HF.

VI.2

But DG equals A and GE to B, and DH to C, therefore A is to B as C is to HF.

V.7

Therefore a fourth proportional HF has been found to the three given straight lines A, B, and C.

Q.E.F.

Proposition 13

To find a mean proportional to two given straight lines.

Let AB and BC be the two given straight lines.
It is required to find a mean proportional to AB and BC.

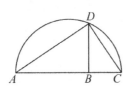

Place them in a straight line, and describe the semicircle ADC on AC. Draw BD from the point B at right angles to the straight line AC, and join AD and DC.

I.11

Since the angle ADC is an angle in a semicircle, it is right.

III.31

And, since, in the right-angled triangle ADC, BD has been drawn from the right angle perpendicular to the base, therefore BD is a mean proportional between the segments of the base, AB and BC.

VI.8,Cor

Therefore a mean proportional BD has been found to the two given straight lines AB and BC.

Q.E.F.

Exercises on Book VI

1. Prove I.43 using tools of Book VI. (Use no subtraction whatever.)

2. Demonstrate two different ways of bisecting a finite straight line. (One is from Book I and one is from Book VI.)

3. Take a straightedge and on it mark (if it is not already marked, as a ruler would be) a line approximately 6 inches long. Take a piece of lined notebook paper and, using your straightedge, compass, and pencil, draw a line approximately six inches long that is divided into seven equal parts. (There is more than one way to do this, but one of them is particularly easy.)

4. Given two parallelograms with equal angles, prove that if the parallelograms are equal then the sides about the equal angles are proportional (reciprocally). Set up your proof as follows:

Suppose AB and BC are equal parallelograms with equal angles. Place them adjacent to one another with point B touching, and in such a manner that FBG and DBE are straight lines. (How do you know you can do this?)

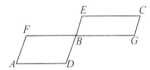

Now prove that $DB:BE = GB:BF$

(Hint: you may have to add a couple of lines to the drawing. Ask your teacher how to get started if you need to.)

5. Prove the converse of Exercise 4. That is, prove that equiangular parallelograms with sides reciprocally proportional are equal. Obviously you can use the same drawing as Exercise 4.

6. Suppose you have two triangles with one angle of one triangle equal to one angle of the other. State a proposition analogous to the proposition of Exercise 4, above. Check with your teacher to make sure you have theorized correctly, then prove your proposition.

7. Again suppose you have two triangles with one angle of one triangle equal to one angle of the other triangle. State and prove the converse of the proposition of Exercise 6.

Exercises 8 through 10 are statements of commonly known and frequently used theorems of plane geometry. Write out a complete proof of these theorems. Outlines for exercises 8 and 10 are provided which you should follow. You should also, of course, provide reasons for statements 1, 2, and 3 in problem 8.

8. Proposition: If four straight lines are proportional, the rectangle contained by the means equals the rectangle contained by the extremes.

Given: Proportional lines *AB*, *CD*, *E*, and *F*. That is, line *AB* is to line *CD* as line *E* is to line *F*

To Prove: The rectangle contained by *AB* and *F* is equal to the rectangle contained by *CD* and *E*.

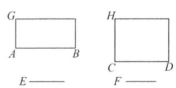

1. Draw *AG* ⊥ *AB* with *AG* = *F*
2. Draw *CH* ⊥ *CD* with *CH* = *E*
3. Complete the parallelograms

We wish to prove that rectangle *BG* equals rectangle *DH*.

Complete this proof by showing:
I. Parallelograms *BG* and *DH* are equiangular
II. Parallelograms *BG* and *DH* have sides reciprocally proportional
III. The rectangles *BG* and *DH* are equal (your final result)

This proposition has a very common algebraic counterpart. What is it?

9. Proposition: Given four straight lines, if the rectangle contained by the means equals the rectangle contained by the extremes, then the four straight lines are proportional. (This is, of course, the converse of Exercise 8.)

10. Similar triangles are to one another in the duplicate ratio of corresponding sides. (Make sure you fully understand the definition of duplicate ratio before trying to complete this proof. Refer back to V Def. 9 if you need to.)

Given: Similar triangles *ABC* and *DEF* with the angle at *B* equal to the angle at *E* and with *AB*:*BC* = *DE*:*EF*

To Prove: △*ABC* : △*DEF* is the duplicate ratio of *BC*:*EF*.

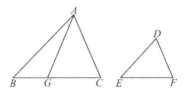

Now in order to talk about the duplicate ratio of *BC:EF* we need a third magnitude having the property that *BC, EF*, and this third magnitude are proportional (V Def. 9).
But, according to VI.11, it is possible to construct such a magnitude.

So, let us construct that third magnitude and let *BG* be drawn on *BC* equal to this new magnitude, the third proportional to *BC* and *EF*. That is, draw *BG* so that *BC:EF = EF:BG*.

Then, according to V Def. 9, *BC:BG* is the duplicate ratio of *BC:EF*. That being the case, we want to show (given △*ABC* and △*DEF* are similar), that △*ABC* : △*DEF* = *BC:BG*.

Complete this proof by proving the following:
I. *AB:DE = BC:EF*
II. *AB:DE = EF:BG*
III. △*ABG* = △*DEF*
IV. △*ABC* : △*DEF* = *BC:BG*

Note: These are not all one step proofs. Also, at some point you may have to make use of the result proved in Exercise 6, above.

Appendix A

Definitions

1 Any rectangular parallelogram is said to be *contained* by the two straight lines containing the right angle.

2 And in any parallelogrammic area let any one whatever of the parallelograms about its diameter with the two complements be called a *gnomon*.

Book II Propositions

Proposition 1

If there are two straight lines, and one of them is cut into any number of segments whatever, then the rectangle contained by the two straight lines equals the sum of the rectangles contained by the uncut straight line and each of the segments.

Let A and BC be two straight lines, and let BC be cut at random at the points D and E.
I say that the rectangle A by BC equals the sum of the rectangle A by BD, the rectangle A by DE, and the rectangle A by EC.

Draw BF from B at right angles to BC. Make BG equal to A. Draw GH through G parallel to BC. Through D, E, and C draw DK, EL, and CH parallel to BG.   I.11 / I.3 / I.31

Then BH equals the sum of BK, DL, and EH.

Now BH is the rectangle A by BC, for it is contained by GB and BC, and BG equals A; BK is the rectangle A by BD, for it is contained by GB and BD, and BG equals A; and DL is the rectangle A by DE, for DK, that is BG, equals A. Similarly also EH is the rectangle A by EC.   II.Def.1 / I.34

Therefore the rectangle A by BC equals the sum of the rectangle A by BD, the rectangle A by DE, and the rectangle A by EC.

Therefore *if there are two straight lines, and one of them is cut into any number of segments whatever, then the rectangle contained by the two straight lines equals the sum of the rectangles contained by the uncut straight line and each of the segments.*

Q.E.D.

Proposition 2

If a straight line is cut at random, then the sum of the rectangles contained by the whole and each of the segments equals the square on the whole.

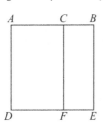

Let the straight line AB be cut at random at the point C.
I say that the sum of the rectangle BA by AC and the rectangle AB by BC equals the square on AB

Describe the square $ADEB$ on AB, and draw CF through C parallel to either AD or BE. Then AE equals AF plus CE.   I.46
                                                                                                                                  I.31

Now AE is the square on AB; AF is the rectangle BA by AC, for it is contained by DA and AC, and AD equals AB; and CE is the rectangle AB by BC, for BE equals AB.   II.Def.1

Therefore the sum of the rectangle BA by AC and the rectangle AB by BC equals the square on AB.

Therefore *if a straight line is cut at random, then the sum of the rectangles contained by the whole and each of the segments equals the square on the whole.*

Q.E.D.

Proposition 3

If a straight line is cut at random, then the rectangle contained by the whole and one of the segments equals the sum of the rectangle contained by the segments and the square on the aforesaid segment.

Let the straight line AB be cut at random at C.

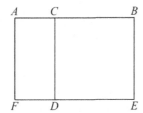

I say that the rectangle AB by BC equals the sum of the rectangle AC by CB and the square on BC.

Describe the square $CDEB$ on CB. Draw ED through to F, and draw AF through A parallel to either CD or BE.   I.46
                                                                                                                          I.31

Then AE equals AD plus CE.

Now AE is the rectangle AB by BC, for it is contained by AB and BE, and BE equals BC; AD is the rectangle AC by CB, for DC equals CB; and DB is the square on CB.

Therefore the rectangle AB by BC equals the sum of the rectangle AC by CB and the square on BC.

Therefore *if a straight line is cut at random, then the rectangle contained by the whole and one of the segments equals the sum of the rectangle contained by the segments and the square on the aforesaid segment.*

Q.E.D.

Proposition 4

If a straight line is cut at random, then the square on the whole equals the sum of the squares on the segments plus twice the rectangle contained by the segments.

Let the straight line AB be cut at random at C.

I say that the square on AB equals the sum of the squares on AC and CB plus twice the rectangle AC by CB.

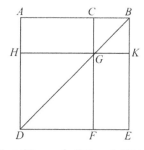

Describe the square ADEB on AB. Join BD. Draw CF through C parallel to either AD or EB, and draw HK through G parallel to either AB or DE. I.46 / I.31

Then, since CF is parallel to AD, and BD falls on them, the exterior angle CGB equals the interior and opposite angle ADB. I.29

But the angle ADB equals the angle ABD, since the side BA also equals AD. Therefore the angle CGB also equals the angle GBC, so that the side BC also equals the side CG. I.5 / I.6

But CB equals GK, and CG to KB. Therefore GK also equals KB. Therefore CGKB is equilateral. I.34

I say next that it is also right-angled.

Since CG is parallel to BK, the sum of the angles KBC and GCB equals two right angles. I.29

But the angle KBC is right. Therefore the angle BCG is also right, so that the opposite angles CGK and GKB are also right. I.34

Therefore CGKB is right-angled, and it was also proved equilateral, therefore it is a square, and it is described on CB.

For the same reason HF is also a square, and it is described on HG, that is AC. Therefore the squares HF and KC are the squares on AC and CB. I.34

Now, since AG equals GE, and AG is the rectangle AC by CB, for GC equals CB, therefore GE also equals the rectangle AC by CB. Therefore the sum of AG and GE equals twice the rectangle AC by CB. I.43

But the squares HF and CK are also the squares on AC and CB, therefore the sum of the four figures HF, CK, AG, and GE equals the sum of the squares on AC and CB plus twice the rectangle AC by CB.

But HF, CK, AG, and GE are the whole ADEB, which is the square on AB.

Therefore the square on AB equals the the sum of the squares on AC and CB plus twice the rectangle AC by CB.

Therefore *if a straight line is cut at random, the square on the whole equals the squares on the segments plus twice the rectangle contained by the segments.*

Q.E.D.

Proposition 5

If a straight line is cut into equal and unequal segments, then the rectangle contained by the unequal segments of the whole together with the square on the straight line between the points of section equals the square on the half.

Let a straight line AB be cut into equal segments at C and into unequal segments at D.
I say that the rectangle AD by DB together with the square on CD equals the square on CB.

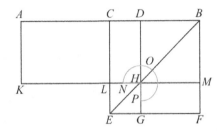

Describe the square $CEFB$ on CB, and join BE. Draw DG through D parallel to either CE or BF, again draw KM through H parallel to either AB or EF, and again draw AK through A parallel to either CL or BM.

I.46

I.31

Then, since the complement CH equals the complement HF, add DM to each. Therefore the whole CM equals the whole DF.

I.43

But CM equals AL, since AC is also equal to CB. Therefore AL also equals DF. Add CH to each. Therefore the whole AH equals the gnomon NOP.

I.36
II.Def.2

But AH is the rectangle AD by DB, for DH equals DB, therefore the gnomon NOP also equals the rectangle AD by DB.

Add LG, which equals the square on CD, to each. Therefore the sum of the gnomon NOP and LG equals the sum of the rectangle AD by DB and the square on CD.

But the gnomon NOP together with LG is the whole square $CEFB$, which is described on CB.

Therefore the rectangle AD by DB together with the square on CD equals the square on CB.

Therefore *if a straight line is cut into equal and unequal segments, then the rectangle contained by the unequal segments of the whole together with the square on the straight line between the points of section equals the square on the half.*

Q.E.D.

Proposition 6

If a straight line is bisected and a straight line is added to it in a straight line, then the rectangle contained by the whole with the added straight line and the added straight line together with the square on the half equals the square on the straight line made up of the half and the added straight line.

Let a straight line AB be bisected at the point C, and let a straight line BD be added to it in a straight line.
I say that the rectangle AD by DB together with the square on CB equals the square on CD.

	Describe the square CEFD on CD, and join DE. Draw BG through the point B parallel to either EC or DF, draw KM through the point H parallel to either AB or EF, and further draw AK through A parallel to either CL or DM.	I.46
		I.31
	Then, since AC equals CB, AL also equals CH. But CH equals HF. Therefore AL also equals HF.	I.36
		I.43

Add CM to each. Therefore the whole AM equals the gnomon NOP.   II.Def.2

But AM is the rectangle AD by DB, for DM equals DB. Therefore the gnomon NOP also equals the rectangle AD by DB.

Add LG, which equals the square on BC, to each. Therefore the rectangle AD by DB together with the square on CB equals the gnomon NOP plus LG.

But the gnomon NOP and LG are the whole square CEFD, which is described on CD.

Therefore the rectangle AD by DB together with the square on CB equals the square on CD.

Therefore *if a straight line is bisected and a straight line is added to it in a straight line, then the rectangle contained by the whole with the added straight line and the added straight line together with the square on the half equals the square on the straight line made up of the half and the added straight line.*

Q.E.D.

Proposition 7

If a straight line is cut at random, then the sum of the square on the whole and that on one of the segments equals twice the rectangle contained by the whole and the said segment plus the square on the remaining segment.

Let a straight line AB be cut at random at the point C.

I say that the sum of the squares on AB and BC equals twice the rectangle AB by BC plus the square on CA.

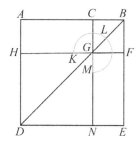

Describe the square ADEB on AB, and let the figure be drawn.	I.46
	I.31
Then, since AG equals GE, add CF to each, therefore the whole AF equals the whole CE.	I.43

Therefore the sum of AF and CE is double AF.

But the sum of AF and CE equals the gnomon KLM plus the square CF, therefore the gnomon KLM plus the square CF is double AF.

But twice the rectangle *AB* by *BC* is also double *AF*, for *BF* equals *BC*, therefore the gnomon *KLM* plus the square *CF* equal twice the rectangle *AB* by *BC*.

Add *DG*, which is the square on *AC*, to each. Therefore the gnomon *KLM* plus the sum of the squares *BG* and *GD* equals twice the rectangle *AB* by *BC* plus the square on *AC*.

But the gnomon *KLM* plus the sum of the squares *BG* and *GD* equals the whole *ADEB* plus *CF*, which are squares described on *AB* and *BC*.

Therefore the sum of the squares on *AB* and *BC* equals twice the rectangle *AB* by *BC* plus the square on *CA*.

Therefore *if a straight line is cut at random, then the sum of the square on the whole and that on one of the segments equals twice the rectangle contained by the whole and the said segment plus the square on the remaining segment.*

Q.E.D.

Proposition 8

If a straight line is cut at random, then four times the rectangle contained by the whole and one of the segments plus the square on the remaining segment equals the square described on the whole and the aforesaid segment as on one straight line.

Let a straight line *AB* be cut at random at the point *C*.
I say that four times the rectangle *AB* by *BC* plus the square on *AC* equals the square described on *AB* and *BC* as on one straight line.

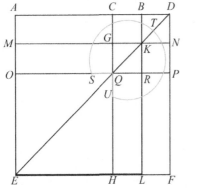

Produce the straight line *BD* in a straight line with *AB*, and make *BD* equal to *CB*. Describe the square *AEFD* on *AD*, and let the figure be drawn. I.3 I.46 I.31

Then, since *CB* equals *BD*, while *CB* equals *GK*, and *BD* equals *KN*, therefore *GK* also equals *KN*. I.34

For the same reason *QR* also equals *RP*.

And, since *BC* equals *BD*, and *GK* equals *KN*, therefore *CK* also equals *KD*, and *GR* equals *RN*. I.36

But *CK* equals *RN*, for they are complements of the parallelogram *CP*.

Therefore *KD* also equals *GR*. Therefore the four areas *DK*, *CK*, *GR*, *RN* equal one another. Therefore the four are quadruple of *CK*. I.43

Again, since *CB* equals *BD*, while *BD* equals *BK*, that is *CG*, and *CB* equals *GK*, that is *GQ*, therefore *CG* also equals *GQ*. I.34

And, since *CG* equals *GQ*, and *QR* equals *RP*, *AG* also equals *MQ*, and *QL* equals *RF*. I.36

But *MQ* equals *QL*, for they are complements of the parallelogram *ML*, therefore *AG* also equals *RF*. Therefore the four areas *AG*, *MQ*, *QL*, *RF* equal one another. Therefore the four are quadruple of *AG*. But the four areas *CK*, *KD*, *GR*, *RN* were proved to be quadruple of *CK*, therefore the eight areas, which contain the gnomon *STU*, are quadruple of *AK*. I.43

Now, since AK is the rectangle AB by BD, for BK equals BD, therefore four times the rectangle AB by BD is quadruple of AK.

But the gnomon STU was also proved to be quadruple of AK, therefore four times the rectangle AB by BD equals the gnomon STU.

Add OH, which equals the square on AC, to each. Therefore four times the rectangle AB by BD plus the square on AC equals the gnomon STU plus OH.

But the gnomon STU and OH are the whole square AEFD, which is described on AD. Therefore four times the rectangle AB by BD plus the square on AC equals the square on AD.

But BD equals BC.

Therefore four times the rectangle AB by BC together with the square on AC equals the square on AD, that is to the square described on AB and BC as on one straight line.

Therefore *if a straight line is cut at random, then four times the rectangle contained by the whole and one of the segments plus the square on the remaining segment equals the square described on the whole and the aforesaid segment as on one straight line.*

Q.E.D.

Proposition 9

If a straight line is cut into equal and unequal segments, then the sum of the squares on the unequal segments of the whole is double the sum of the square on the half and the square on the straight line between the points of section.

Let a straight line AB be cut into equal segments at C, and into unequal segments at D. I say that the sum of the squares on AD and DB is double the sum of the squares on AC and CD.

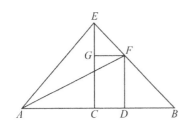

Draw CE from C at right angles to AB, and make it equal to either AC or CB. Join EA and EB. Draw DF through D parallel to EC and FG through F parallel to AB. Join AF.

I.11
I.3
I.31

Then, since AC equals CE, the angle EAC also equals the angle AEC.

I.5

And, since the angle at C is right, the sum of the remaining angles EAC and AEC equals one right angle.

I.32

And they are equal, therefore each of the angles CEA and CAE is half of a right angle. For the same reason each of the angles CEB and EBC is also half of a right angle, therefore the whole angle AEB is right.

And, since the angle GEF is half of a right angle, and the angle EGF is right, for it equals the interior and opposite angle ECB, the remaining angle EFG is half of a right angle. Therefore the angle GEF equals the angle EFG, so that the side EG also equals GF.

I.29
I.32
I.6

Again, since the angle at B is half of a right angle, and the angle FDB is right, for it is again equal to the interior and opposite angle ECB, the remaining angle BFD is half of a right angle. Therefore the angle at B equals the angle DFB, so that the side FD also equals the side DB.

I.29
I.32
I.6

Now, since AC equals CE, the square on AC also equals the square on CE, therefore the sum of the squares on AC and CE is double the square on AC.

But the square on EA equals the sum of the squares on AC and CE, for the angle ACE is right, therefore the square on EA is double the square on AC.   I.47

Again, since EG equals GF, the square on EG also equals the square on GF. Therefore the sum of the squares on EG and GF is double the square on GF.
But the square on EF equals the sum of the squares on EG and GF, therefore the square on EF is double the square on GF.   I.47

But GF equals CD, therefore the square on EF is double the square on CD.   I.34

But the square on EA is also double of the square on AC, therefore the sum of the squares on AE and EF is double the sum of the squares on AC and CD.

And the square on AF equals sum of the squares on AE and EF, for the angle AEF is right. Therefore the square on AF is double the sum of the squares on AC and CD.   I.47

But the sum of the squares on AD and DF equals the square on AF, for the angle at D is right, therefore the sum of the squares on AD and DF is double the sum the squares on AC and CD.   I.47

And DF equals DB.
Therefore the sum of the squares on AD and DB is double the sum of the squares on AC and CD.

Therefore *if a straight line is cut into equal and unequal segments, then the sum of the squares on the unequal segments of the whole is double the sum of the square on the half and the square on the straight line between the points of section.*

Q.E.D.

Proposition 10

If a straight line is bisected, and a straight line is added to it in a straight line, then the square on the whole with the added straight line and the square on the added straight line both together are double the sum of the square on the half and the square described on the straight line made up of the half and the added straight line as on one straight line.

Let a straight line AB be bisected at C, and let a straight line BD be added to it in a straight line.
I say that the sum of the squares on AD and DB is double the sum of the squares on AC and CD.

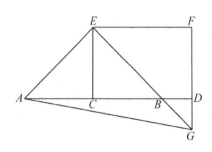

Draw CE from the point C at right angles to AB and equal to either AC or CB. Join EA and EB. Draw EF through E parallel to AD, and draw FD through D parallel to CE.   I.11 / I.3 / I.31

Then, since a straight line EF falls on the parallel straight lines EC and FD, the sum of the angles CEF and EFD equals two right angles. Therefore the sum of the angles FEB and EFD is less than two right angles.   I.29

But straight lines produced from angles less than two right angles meet. Therefore *EB* and *FD*, if produced in the direction *B* and *D*, will meet.   Post.5

Let them be produced and meet at *G*, and join *AG*.

Then, since *AC* equals *CE*, the angle *EAC* also equals the angle *AEC*. The angle at *C* is right, therefore each of the angles *EAC* and *AEC* is half of a right angle.   I.5
I.32

For the same reason each of the angles *CEB* and *EBC* is also half of a right angle, therefore the angle *AEB* is right.

And, since the angle *EBC* is half of a right angle, the angle *DBG* is also half of a right angle. But the angle *BDG* is also right, for it equals the angle *DCE*, since they are alternate. Therefore the remaining angle *DGB* is half of a right angle. Therefore the angle *DGB* equals the angle *DBG*, so that the side *BD* also equals the side *GD*.   I.15
I.29
I.32
I.6

Again, since the angle *EGF* is half of a right angle, and the angle at *F* is right, for it equals the opposite angle, the angle at *C*, the remaining angle *FEG* is half of a right angle. Therefore the angle *EGF* equals the angle *FEG*, so that the side *GF* also equals the side *EF*.   I.34
I.32
I.6

Now, since the square on *EC* equals the square on *CA*, the sum of the squares on *EC* and *CA* is double the square on *CA*. But the square on *EA* equals the sum of the squares on *EC* and *CA*, therefore the square on *EA* is double the square on *AC*.   I.47

Again, since *FG* equals *EF*, the square on *FG* also equals the square on *FE*. Therefore the sum of the squares on *GF* and *FE* is double the square on *EF*. But the square on *EG* equals the sum of the squares on *GF* and *FE*, therefore the square on *EG* is double the square on *EF*.   I.47

And *EF* equals *CD*, therefore the square on *EG* is double the square on *CD*. But the square on *EA* was also proved to be double the square on *AC*, therefore the sum of the squares on *AE* and *EG* is double the sum of the squares on *AC* and *CD*.   I.34

And the square on *AG* equals the sum of the squares on *AE* and *EG*, therefore the square on *AG* is double the sum of the squares on *AC* and *CD*. But the sum of the squares on *AD* and *DG* equals the square on *AG*, therefore the sum of the squares on *AD* and *DG* is double the sum of the squares on *AC* and *CD*.   I.47

And *DG* equals *DB*, therefore the sum of the squares on *AD* and *DB* is double the sum of the squares on *AC* and *CD*.

Therefore *if a straight line is bisected, and a straight line is added to it in a straight line, then the square on the whole with the added straight line and the square on the added straight line both together are double the sum of the square on the half and the square described on the straight line made up of the half and the added straight line as on one straight line.*

Q.E.D.

Proposition 11

To cut a given straight line so that the rectangle contained by the whole and one of the segments equals the square on the remaining segment.

Let AB be the given straight line.

It is required to cut AB so that the rectangle contained by the whole and one of the segments equals the square on the remaining segment.

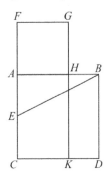

Describe the square ABDC on AB. Bisect AC at the point E, and join BE. Draw CA through to F, and make EF equal to BE. Describe the square FH on AF, and draw GH through to K.

I.46
I.10
I.3
I.46

I say that AB has been cut at H so that the rectangle AB by BH equals the square on AH.

Since the straight line AC has been bisected at E, and FA is added to it, the rectangle CF by FA together with the square on AE equals the square on EF.

II.6

But EF equals EB, therefore the rectangle CF by FA together with the square on AE equals the square on EB.

But the sum of the squares on BA and AE equals the square on EB, for the angle at A is right, therefore the rectangle CF by FA together with the square on AE equals the sum of the squares on BA and AE.

I.47

Subtract the square on AE from each. Therefore the remaining rectangle CF by FA equals the square on AB.

Now the rectangle CF by FA is FK, for AF equals FG, and the square on AB is AD, therefore FK equals AD.

Subtract AK from each. Therefore FH, which remains, equals HD.

And HD is the rectangle AB by BH, for AB equals BD, and FH is the square on AH, therefore the rectangle AB by BH equals the square on HA.

Therefore the given straight line AB has been cut at H so that the rectangle AB by BH equals the square on HA.

Q.E.F.

Proposition 12

In obtuse-angled triangles the square on the side opposite the obtuse angle is greater than the sum of the squares on the sides containing the obtuse angle by twice the rectangle contained by one of the sides about the obtuse angle, namely that on which the perpendicular falls, and the straight line cut off outside by the perpendicular towards the obtuse angle.

Let ABC be an obtuse-angled triangle having the angle BAC obtuse, and draw BD from the point B perpendicular to CA produced.

I.12

I say that the square on BC is greater than the sum of the squares on BA and AC by twice the rectangle CA by AD.

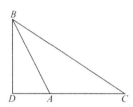

Since the straight line *CD* has been cut at random at the point *A*, the square on *DC* equals the sum of the squares on *CA* and *AD* and twice the rectangle *CA* by *AD*.

II.4

Add the square on *DB* be added to each. Therefore the sum of the squares on *CD* and *DB* equals the sum of the squares on *CA*, *AD*, and *DB* plus twice the rectangle *CA* by *AD*.

But the square on *CB* equals the sum of the squares on *CD* and *DB*, for the angle at *D* is right, and the square on *AB* equals the sum of the squares on *AD* and *DB*, therefore the square on *CB* equals the sum of the squares on *CA* and *AB* plus twice the rectangle *CA* by *AD*, so that the square on *CB* is greater than the sum of the squares on *CA* and *AB* by twice the rectangle *CA* by *AD*.

I.47

Therefore *in obtuse-angled triangles the square on the side opposite the obtuse angle is greater than the sum of the squares on the sides containing the obtuse angle by twice the rectangle contained by one of the sides about the obtuse angle, namely that on which the perpendicular falls, and the straight line cut off outside by the perpendicular towards the obtuse angle.*

Q.E.D.

Proposition 13

In acute-angled triangles the square on the side opposite the acute angle is less than the sum of the squares on the sides containing the acute angle by twice the rectangle contained by one of the sides about the acute angle, namely that on which the perpendicular falls, and the straight line cut off within by the perpendicular towards the acute angle.

Let *ABC* be a triangle having the angle at *B* acute, and draw *AD* from the point *A* perpendicular to *BC*.

I.12

I say that the square on *AC* is less than the sum of the squares on *CB* and *BA* by twice the rectangle *CB* by *BD*.

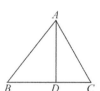

Since the straight line *CB* has been cut at random at *D*, the sum of the squares on *CB* and *BD* equals twice the rectangle *CB* by *BD* plus the square on *DC*.

II.7

Add the square on *DA* to each. Therefore the sum of the squares on *CB*, *BD*, and *DA* equals twice the rectangle *CB* by *BD* plus the sum of the squares on *AD* and *DC*.

But the square on *AB* equals the sum of the squares on *BD* and *DA*, for the angle at *D* is right, and the square on *AC* equals the sum of the squares on *AD* and *DC*, therefore the sum of the squares on *CB* and *BA* equals the square on *AC* plus twice the rectangle *CB* by *BD*, so that the square on *AC* alone is less than the sum of the squares on *CB* and *BA* by twice the rectangle *CB* by *BD*.

I.47

Proposition 14

To construct a square equal to a given rectilinear figure.

Let *A* be the given rectilinear figure.

It is required to construct a square equal to the rectilinear figure *A*.

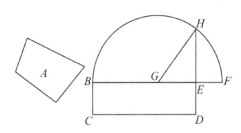

Construct the rectangular parallelogram *BD* equal to the rectilinear figure *A*. — I.45

Then, if *BE* equals *ED*, then that which was proposed is done, for a square *BD* has been constructed equal to the rectilinear figure *A*. But, if not, one of the straight lines *BE* or *ED* is greater.

Let *BE* be greater, and produce it to *F*. Make *EF* equal to *ED*, and bisect *BF* at *G*. — I.3, I.10

Describe the semicircle *BHF* with center *G* and radius one of the straight lines *GB* or *GF*. Produce *DE* to *H*, and join *GH*. — I.Def.18

Then, since the straight line *BF* has been cut into equal segments at *G* and into unequal segments at *E*, the rectangle *BE* by *EF* together with the square on *EG* equals the square on *GF*. — II.5

But *GF* equals *GH*, therefore the rectangle *BE* by *EF* together with the square on *GE* equals the square on *GH*.

But the sum of the squares on *HE* and *EG* equals the square on *GH*, therefore the rectangle *BE* by *EF* together with the square on *GE* equals the sum of the squares on *HE* and *EG*. — I.47

Subtract the square on *GE* from each. Therefore the remaining rectangle *BE* by *EF* equals the square on *EH*.

But the rectangle *BE* by *EF* is *BD*, for *EF* equals *ED*, therefore the parallelogram *BD* equals the square on *HE*.

And *BD* equals the rectilinear figure *A*.

Therefore the rectilinear figure *A* also equals the square which can be described on *EH*.

Therefore a square, namely that which can be described on *EH*, has been constructed equal to the given rectilinear figure *A*.

Q.E.F.

Book III Propositions

Proposition 35

If in a circle two straight lines cut one another, then the rectangle contained by the segments of the one equals the rectangle contained by the segments of the other.

For in the circle ABCD let the two straight lines AC and BD cut one another at the point E.

I say that the rectangle AE by EC equals the rectangle DE by EB.

If now AC and BD are through the center, so that E is the center of the circle ABCD, it is manifest that, AE, EC, DE, and EB being equal, the rectangle AE by EC also equals the rectangle DE by EB.

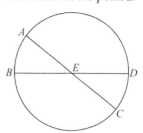

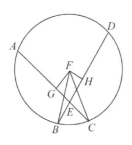

Next let AC and DB not be through the center. Take the center F of ABCD. Draw FG and FH from F perpendicular to the straight lines AC and DB. Join FB, FC, and FE.

III.1
I.12

Then, since a straight line GF through the center cuts a straight line AC not through the center at right angles, it also bisects it, therefore AG equals GC.

III.3

Since, then, the straight line AC has been cut into equal parts at G and into unequal parts at E, the rectangle AE by EC together with the square on EG equals the square on GC.

II.5

Add the square on GF. Therefore the rectangle AE by EC plus the sum of the squares on GE and GF equals the sum of the squares on CG and GF.

But the square on FE equals the sum of the squares on EG and GF, and the square on FC equals the sum of the squares on CG and GF. Therefore the rectangle AE by EC plus the square on FE equals the square on FC.

And FC equals FB, therefore the rectangle AE by EC plus the square on EF equals the square on FB.

I.47

For the same reason, also, the rectangle DE by EB plus the square on FE equals the square on FB.

But the rectangle AE by EC plus the square on FE was also proved equal to the square on FB, therefore the rectangle AE by EC plus the square on FE equals the rectangle DE by EB plus the square on FE.

Subtract the square on FE from each. Therefore the remaining rectangle AE by EC equals the rectangle DE by EB.

Therefore *if in a circle two straight lines cut one another, then the rectangle contained by the segments of the one equals the rectangle contained by the segments of the other.*

Q.E.D.

Proposition 36

If a point is taken outside a circle and two straight lines fall from it on the circle, and if one of them cuts the circle and the other touches it, then the rectangle contained by the whole of the straight line which cuts the circle and the straight line intercepted on it outside between the point and the convex circumference equals the square on the tangent.

Let a point D be taken outside the circle ABC, and from D let the two straight lines DCA and DB fall on the circle ABC. Let DCA cut the circle ABC, and let BD touch it. I say that the rectangle AD by DC equals the square on DB.

Then DCA is either through the center or not through the center.

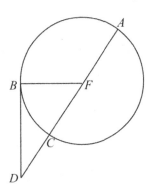

First let it be through the center, and let F be the center of the circle ABC. Join FB. Therefore the angle FBD is right.   III.18

And, since AC has been bisected at F, and CD is added to it, the rectangle AD by DC plus the square on FC equals the square on FD.   II.6

But FC equals FB, therefore the rectangle AD by DC plus the square on FB equals the square on FD.

And the sum of the squares on FB and BD equals the square on FD, therefore the rectangle AD by DC plus the square on FB equals the sum of the squares on FB and BD.   I.47

Subtract the square on FB from each. Therefore the remaining rectangle AD by DC equals the square on the tangent DB.

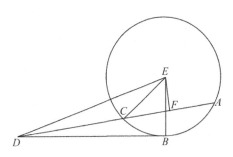

Again, let DCA not be through the center of the circle ABC. Take the center E, and draw EF from E perpendicular to AC. Join EB, EC, and ED.   III.1

Then the angle EBD is right.   III.18

And, since a straight line EF through the center cuts a straight line AC not through the center at right angles, it also bisects it, therefore AF equals FC.   III.3

Now, since the straight line AC has been bisected at the point F, and CD is added to it, the rectangle AD by DC plus the square on FC equals the square on FD. II.6

Add the square on FE to each. Therefore the rectangle AD by DC plus the sum of the squares on CF and FE equals the sum of the squares on FD and FE.

But the square on EC equals the sum of the squares on CF and FE, for the angle EFC is right, and the square on ED equals the sum of the squares on DF and FE, therefore the rectangle AD by DC plus the square on EC equals the square on ED. I.47

And EC equals EB, therefore the rectangle AD by DC plus the square on EB equals the square on ED.

But the sum of the squares on EB and BD equals the square on ED, for the angle EBD is right, therefore the rectangle AD by DC plus the square on EB equals the sum of the squares on EB and BD. I.47

Subtract the square on EB from each. Therefore the remaining rectangle AD by DC equals the square on DB.

Therefore *if a point is taken outside a circle and two straight lines fall from it on the circle, and if one of them cuts the circle and the other touches it, then the rectangle contained by the whole of the straight line which cuts the circle and the straight line intercepted on it outside between the point and the convex circumference equals the square on the tangent.*

Q.E.D.

Proposition 37

If a point is taken outside a circle and from the point there fall on the circle two straight lines, if one of them cuts the circle, and the other falls on it, and if further the rectangle contained by the whole of the straight line which cuts the circle and the straight line intercepted on it outside between the point and the convex circumference equals the square on the straight line which falls on the circle, then the straight line which falls on it touches the circle.

Let a point D be taken outside the circle ABC, from D let the two straight lines DCA and DB fall on the circle ACB, let DCA cut the circle and DB fall on it, and let the rectangle AD by DC equal the square on DB.
I say that DB touches the circle ABC.

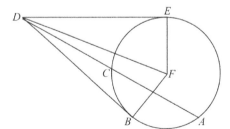

Draw DE touching ABC. Take the center F of the circle ABC, and join FE, FB, and FD. III.17, III.1

Thus the angle FED is right. III.18

Now, since DE touches the circle ABC, and DCA cuts it, the rectangle AD by DC equals the square on DE. III.36

But the rectangle *AD* by *DC* was also equal to the square on *DB*, therefore the square on *DE* equals the square on *DB*. Therefore *DE* equals *DB*.

And *FE* equals *FB*, therefore the two sides *DE* and *EF* equal the two sides *DB* and *BF*, and *FD* is the common base of the triangles, therefore the angle *DEF* equals the angle *DBF*.   I.8

But the angle *DEF* is right, therefore the angle *DBF* is also right.

And *FB* produced is a diameter, and the straight line drawn at right angles to the diameter of a circle, from its end, touches the circle, therefore *DB* touches the circle.   III.16 Cor

Similarly this can be proved to be the case even if the center is on *AC*.

Therefore *if a point is taken outside a circle and from the point there fall on the circle two straight lines, if one of them cuts the circle, and the other falls on it, and if further the rectangle contained by the whole of the straight line which cuts the circle and the straight line intercepted on it outside between the point and the convex circumference equals the square on the straight line which falls on the circle, then the straight line which falls on it touches the circle.*

Q.E.D.

Book IV Propositions

Proposition 10

To construct an isosceles triangle having each of the angles at the base double the remaining one.

Set out any straight line AB, and cut it at the point C so that the rectangle AB by BC equals the square on CA. Describe the circle BDE with center A and radius AB. Fit in the circle BDE the straight line BD equal to the straight line AC which is not greater than the diameter of the circle BDE.

Join AD and DC, and circumscribe the circle ACD about the triangle ACD.

II.11
IV.1

IV.5

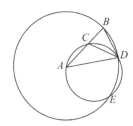

Then, since the rectangle AB by BC equals the square on AC, and AC equals BD, therefore the rectangle AB by BC equals the square on BD.

And, since a point B was taken outside the circle ACD, and from B the two straight lines BA and BD fall on the circle ACD, and one of them cuts it while the other falls on it, and the rectangle AB by BC equals the square on BD, therefore BD touches the circle ACD.

III.37

Since, then, BD touches it, and DC is drawn across from the point of contact at D, therefore the angle BDC equals the angle DAC in the alternate segment of the circle.

III.32

Since, then, the angle BDC equals the angle DAC, add the angle CDA to each, therefore the whole angle BDA equals the sum of the two angles CDA and DAC.

But the exterior angle BCD equals the sum of the angles CDA and DAC, therefore the angle BDA also equals the angle BCD.

III.32

But the angle BDA equals the angle CBD, since the side AD also equals AB, [I. 5] so that the angle DBA also equals the angle BCD.

I.5

Therefore the three angles BDA, DBA, and BCD equal one another.

And, since the angle DBC equals the angle BCD, the side BD also equals the side DC.

I.6

But BD equals CA by hypothesis, therefore CA also equals CD, so that the angle CDA also equals the angle DAC. Therefore the sum of the angles CDA and DAC is double the angle DAC.

I.5

And the angle BCD equals the sum of the angles CDA and DAC, therefore the angle BCD is also double the angle CAD.

But the angle BCD equals each of the angles BDA and DBA, therefore each of the angles BDA and DBA is also double the angle DAB.

Therefore the isosceles triangle ABD has been constructed having each of the angles at the base DB double the remaining one.

Q.E.F.

Proposition 11

To inscribe an equilateral and equiangular pentagon in a given circle.

Let *ABCDE* be the given circle.

It is required to inscribe an equilateral and equiangular pentagon in the circle *ABCDE*.

Set out the isosceles triangle *FGH* having each of the angles at *G* and *H* double the angle at *F*. Inscribe in the circle *ABCDE* the triangle *ACD* equiangular with the triangle *FGH*, so that the angles *CAD*, *ACD*, and *CDA* equal the angles at *F*, *G*, and *H* respectively. Therefore each of the angles *ACD* and *CDA* is also double the angle *CAD*.

IV.10
IV.2

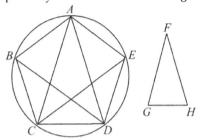

Now bisect the angles *ACD* and *CDA* respectively by the straight lines *CE* and *DB*, and join *AB*, *BC*, *DE*, and *EA*.

I.9

Then, since each of the angles *ACD* and *CDA* is double the angle *CAD*, and they are bisected by the straight lines *CE* and *DB*, therefore the five angles *DAC*, *ACE*, *ECD*, *CDB*, and *BDA* equal one another.

But equal angles stand on equal circumferences, therefore the five circumferences *AB*, *BC*, *CD*, *DE*, and *EA* equal one another.

III.26

But straight lines that cut off equal circumferences are equal, therefore the five straight lines *AB*, *BC*, *CD*, *DE*, and *EA* equal one another. Therefore the pentagon *ABCDE* is equilateral.

III.29

I say next that it is also equiangular.

For, since the circumference *AB* equals the circumference *DE*, add *BCD* to each, therefore the whole circumference *ABCD* equals the whole circumference *EDCB*. And the angle *AED* stands on the circumference *ABCD*, and the angle *BAE* on the circumference *EDCB*, therefore the angle *BAE* also equals the angle *AED*.

III.27

For the same reason each of the angles *ABC*, *BCD*, and *CDE* also equals each of the angles *BAE* and *AED*, therefore the pentagon *ABCDE* is equiangular.

But it was also proved equilateral, therefore an equilateral and equiangular pentagon has been inscribed in the given circle.

Q.E.F.

Proposition 12

To circumscribe an equilateral and equiangular pentagon about a given circle.

Let *ABCDE* be the given circle.

It is required to circumscribe an equilateral and equiangular pentagon about the circle *ABCDE*.

Let *A*, *B*, *C*, *D*, and *E* be conceived to be the angular points of the inscribed pentagon, so that the circumferences *AB*, *BC*, *CD*, *DE*, and *EA* are equal. Draw *GH*, *HK*, *KL*, *LM*, and *MG* through *A*, *B*, *C*, *D*, and *E* touching the circle. Take the center *F* of the circle *ABCDE*, and join *FB*, *FK*, *FC*, *FL*, and *FD*.

IV.11
III.16,Cor
III.1

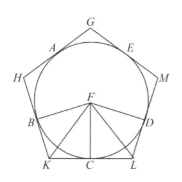

Then, since the straight line *KL* touches the circle *ABCDE* at *C*, and *FC* has been joined from the center *F* to the point of contact at *C*, therefore *FC* is perpendicular to *KL*. Therefore each of the angles at *C* is right. III.18

For the same reason the angles at the points *B* and *D* are also right.

And, since the angle *FCK* is right, therefore the square on *FK* equals the sum of the squares on *FC* and *CK*. I.47

For the same reason the square on *FK* also equals the sum of the squares on *FB* and *BK*, so that the sum of the squares on *FC* and *CK* equals the sum of the squares on *FB* and *BK*, of which the square on *FC* equals the square on *FB*, therefore the remaining square on *CK* equals the square on *BK*. I.47

Therefore *BK* equals *CK*.

And, since *FB* equals *FC*, and *FK* is common, the two sides *BF* and *FK* equal the two sides *CF* and *FK*, and the base *BK* equals the base *CK*, therefore the angle *BFK* equals the angle *KFC*, and the angle *BKF* equals the angle *FKC*. Therefore the angle *BFC* is double the angle *KFC*, and the angle *BKC* double the angle *FKC*. I.8

For the same reason the angle *CFD* is also double the angle *CFL*, and the angle *DLC* double the angle *FLC*.

Now, since the circumference *BC* equals *CD*, the angle *BFC* also equals the angle *CFD*. III.27

And the angle *BFC* is double the angle *KFC*, and the angle *DFC* double the angle *LFC*, therefore the angle *KFC* also equals the angle *LFC*.

But the angle *FCK* also equals the angle *FCL*, therefore *FKC* and *FLC* are two triangles having two angles equal to two angles and one side equal to one side, namely *FC* which is common to them, therefore they will also have the remaining sides equal to the remaining sides, and the remaining angle to the remaining angle, therefore the straight line *KC* equals *CL*, and the angle *FKC* equals the angle *FLC*. I.26

And, since *KC* equals *CL*, therefore *KL* is double *KC*.
For the same reason it can be proved that *HK* is also double *BK*.
And *BK* equals *KC*, therefore *HK* also equals *KL*.
Similarly each of the straight lines *HG*, *GM*, and *ML* can also be proved equal to each of the straight lines *HK* and *KL*, therefore the pentagon *GHKLM* is equilateral.
I say next that it is also equiangular.
For, since the angle *FKC* equals the angle *FLC*, and the angle *HKL* was proved double the angle *FKC*, and the angle *KLM* double the angle *FLC*, therefore the angle *HKL* also equals the angle *KLM*.
Similarly each of the angles *KHG*, *HGM*, and *GML* can also be proved equal to each of the angles *HKL* and *KLM*. Therefore the five angles *GHK*, *HKL*, *KLM*, *LMG*, and *MGH* equal one another.

Therefore the pentagon GHKLM is equiangular.
And it was also proved equilateral, and it has been circumscribed about the circle ABCDE.

Q.E.F.

Proposition 13

To inscribe a circle in a given equilateral and equiangular pentagon.

Let ABCDE be the given equilateral and equiangular pentagon.
It is required to inscribe a circle in the pentagon ABCDE.

Bisect the angles BCD and CDE by the straight lines CF and DF respectively. Join the straight lines FB, FA, and FE from the point F at which the straight lines CF and DF meet one another.   I.9

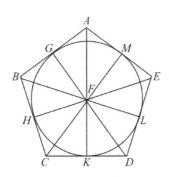

Then, since BC equals CD, and CF common, the two sides BC and CF equal the two sides DC and CF, and the angle BCF equals the angle DCF, therefore the base BF equals the base DF, and the triangle BCF equals the triangle DCF, and the remaining angles equal the remaining angles, namely those opposite the equal sides.   I.4

Therefore the angle CBF equals the angle CDF. And, since the angle CDE is double the angle CDF, and the angle CDE equals the angle ABC, while the angle CDF equals the angle CBF, therefore the angle CBA is also double the angle CBF. Therefore the angle ABF equals the angle FBC. Therefore the angle ABC is bisected by the straight line BF.

Similarly it can be proved that the angles BAE and AED are also bisected by the straight lines FA and FE respectively.

Now draw FG, FH, FK, FL, and FM from the point F perpendicular to the straight lines AB, BC, CD, DE, and EA.   I.12

Then, since the angle HCF equals the angle KCF, and the right angle FHC also equals the angle FKC, FHC and FKC are two triangles having two angles equal to two angles and one side equal to one side, namely FC which is common to them and opposite one of the equal angles, therefore they also have the remaining sides equal to the remaining sides. Therefore the perpendicular FH equals the perpendicular FK.   I.26

©Similarly it can be proved that each of the straight lines FL, FM, and FG also equals each of the straight lines FH and FK, therefore the five straight lines FG, FH, FK, FL, and FM equal one another.

Therefore the circle described with center *F* and radius one of the straight lines *FG*, *FH*, *FK*, *FL*, or *FM* also passes through the remaining points, and it touches the straight lines *AB*, *BC*, *CD*, *DE*, and *EA*, because the angles at the points *G*, *H*, *K*, *L*, and *M* are right.

For, if it does not touch them. but cuts them, it will result that the straight line drawn at right angles to the diameter of the circle from its end falls within the circle, which was proved absurd.

III.16

Therefore the circle described with center *F* and radius one of the straight lines *FG*, *FH*, *FK*, *FL*, or *FM* does not cut the straight lines *AB*, *BC*, *CD*, *DE*, and *EA*. Therefore it touches them.

Let it be described, as *GHKLM*.

Therefore a circle has been inscribed in the given equilateral and equiangular pentagon.

Q.E.F.

Proposition 14

To circumscribe a circle about a given equilateral and equiangular pentagon.

Let *ABCDE* be the given pentagon, which is equilateral and equiangular.
It is required to circumscribe a circle about the pentagon *ABCDE*.

Bisect the angles *BCD* and *CDE* by the straight lines *CF* and *DF* respectively. Join the straight lines *FB*, *FA*, and *FE* from the point *F* at which the straight lines meet to the points *B*, *A*, and *E*.

I.9

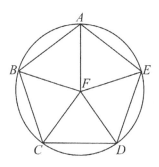

Then in manner similar to the preceding it can be proved that the angles *CBA*, *BAE*, and *AED* are also bisected by the straight lines *FB*, *FA*, and *FE* respectively.

Now, since the angle *BCD* equals the angle *CDE*, and the angle *FCD* is half of the angle *BCD*, and the angle *CDF* half of the angle *CDE*, therefore the angle *FCD* also equals the angle *CDF*, so that the side *FC* also equals the side *FD*.

I.6

Similarly it can be proved that each of the straight lines *FB*, *FA*, and *FE* also equals each of the straight lines *FC* and *FD*. Therefore the five straight lines *FA*, *FB*, *FC*, *FD*, and *FE* equal one another.

Therefore the circle described with center *F* and radius one of the straight lines *FA*, *FB*, *FC*, *FD*, or *FE* also passes through the remaining points, and is circumscribed.
Let it be circumscribed, and let it be *ABCDE*.

Therefore a circle has been circumscribed about the given equilateral and equiangular pentagon.

Q.E.F.

Proposition 15

To inscribe an equilateral and equiangular hexagon in a given circle.

Let *ABCDEF* be the given circle.

It is required to inscribe an equilateral and equiangular hexagon in the circle *ABCDEF*.

Draw the diameter *AD* of the circle *ABCDEF*. Take the center *G* of the circle. Describe the circle *EGCH* with center *D* and radius *DG*. Join *EG* and *CG* and carry them through to the points *B* and *F*. Join *AB*, *BC*, *CD*, *DE*, *EF*, and *FA*.   III.1

I say that the hexagon *ABCDEF* is equilateral and equiangular.

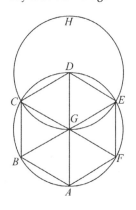

For, since the point *G* is the center of the circle *ABCDEF*, *GE* equals *GD*.

Again, since the point *D* is the center of the circle *GCH*, *DE* equals *DG*.

But *GE* was proved equal to *GD*, therefore *GE* also equals *ED*. Therefore the triangle *EGD* is equilateral, and therefore its three angles *EGD*, *GDE*, and *DEG* equal one another, inasmuch as, in isosceles triangles, the angles at the base equal one another.   I.5

And the sum of the three angles of the triangle equals two right angles, therefore the angle *EGD* is one-third of two right angles.   I.32

Similarly, the angle *DGC* can also be proved to be one-third of two right angles.

And, since the straight line *CG* standing on *EB* makes the sum of the adjacent angles *EGC* and *CGB* equal to two right angles, therefore the remaining angle *CGB* is also one-third of two right angles.   I.13

Therefore the angles *EGD*, *DGC*, and *CGB* equal one another, so that the angles vertical to them, the angles *BGA*, *AGF*, and *FGE*, are equal.   I.15

Therefore the six angles *EGD*, *DGC*, *CGB*, *BGA*, *AGF*, and *FGE* equal one another.

But equal angles stand on equal circumferences, therefore the six circumferences *AB*, *BC*, *CD*, *DE*, *EF*, and *FA* equal one another.   III.26

And straight lines that cut off equal circumferences are equal, therefore the six straight lines equal one another. Therefore the hexagon *ABCDEF* is equilateral.   III.29

I say next that it is also equiangular.

For, since the circumference *FA* equals the circumference *ED*, add the circumference *ABCD* to each, therefore the whole *FABCD* equals the whole *EDCBA*. And the angle *FED* stands on the circumference *FABCD*, and the angle *AFE* on the circumference *EDCBA*, therefore the angle *AFE* equals the angle *DEF*.   III.27

Similarly it can be proved that the remaining angles of the hexagon ABCDEF are also severally equal to each of the angles AFE and FED, therefore the hexagon ABCDEF is equiangular.

But it was also proved equilateral, and it has been inscribed in the circle ABCDEF. Therefore an equilateral and equiangular hexagon has been inscribed in the given circle.

Q.E.F.

Proposition 16

To inscribe an equilateral and equiangular fifteen-angled figure in a given circle.

Let ABCD be the given circle.

It is required to inscribe in the circle ABCD a fifteen-angled figure which shall be both equilateral and equiangular.

Inscribe a side AC of an equilateral triangle and a side AB of an equilateral pentagon in in the circle ABCD. Therefore, of the equal segments of which there are fifteen in the circle ABCD, there will be five in the circumference ABC which is one-third of the circle, and there will be three in the circumference AB which is one-fifth of the circle. Therefore in the remainder BC there will be two of the equal segments.

IV.2
IV.11

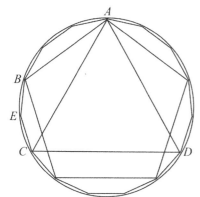

Inscribe a side AC of an equilateral triangle and a side AB of an equilateral pentagon in in the circle ABCD. Therefore, of the equal segments of which there are fifteen in the circle ABCD, there will be five in the circumference ABC which is one-third of the circle, and there will be three in the circumference AB which is one-fifth of the circle. Therefore in the remainder BC there will be two of the equal segments.

IV.2
IV.11

Bisect BC at E. Therefore each of the circumferences BE and EC is a fifteenth of the circle ABCD.

III.30

If therefore we join BE and EC and continually fit into the circle ABCD straight lines equal to them, a fifteen-angled figure which is both equilateral and equiangular will be inscribed in it.

IV.1

Q.E.F.

Book V Propositions

Proposition 23

If there are three magnitudes, and others equal to them in multitude, which taken two and two together are in the same ratio, and the proportion of them be perturbed, then they are also in the same ratio ex aequali.

Let there be three magnitudes A, B, and C, and others D, E, and F, equal to them in multitude, which, taken two and two together, are in the same proportion, and let the proportion of them be perturbed, so that A is to B as E is to F, and B is to C as D is to E.   V.Def.18

I say that A is to C as D is to F.

Take equimultiples G, H, and K of A, B, and D, and take other, arbitrary, equimultiples L, M, and N of C, E, and F.

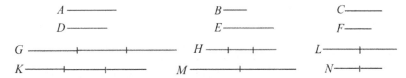

Then, since G and H are equimultiples of A and B, and parts have the same ratio as their multiples, therefore A is to B as G is to H.   V.15

For the same reason E is to F as M is to N. And A is to B as E is to F, therefore G is to H as M is to N.   V.11

Next, since B is to C as D is to E, alternately, also, B is to D as C is to E.   (V.16)

And, since H and K are equimultiples of B and D, and parts have the same ratio as their equimultiples, therefore B is to D as H is to K.   V.15

But B is to D as C is to E, therefore also, H is to K as C is to E.   V.11

Again, since L and M are equimultiples of C and E, therefore C is to E as L is to M.   V.15

But C is to E as H is to K, therefore also, H is to K as L is to M, and, alternately, H is to L as K is to M.   V.11 / (V.16)

But it was also proved that G is to H as M is to N.

Since, then, there are three magnitudes G, H, and L, and others equal to them in multitude K, M, and N, which taken two and two together are in the same ratio, and the proportion of them is perturbed, therefore, ex aequali, if G is in excess of L, K is also in excess of N; if equal, equal; and if less, less.   V.21

And G and K are equimultiples of A and D, and L and N of C and F.

Therefore A is to C as D is to F.   V.Def.5

Therefore, *if there are three magnitudes, and others equal to them in multitude, which taken two and two together are in the same ratio, and the proportion of them be perturbed, then they are also in the same ratio ex aequali.*

Q.E.D.

Proposition 24

If a first magnitude has to a second the same ratio as a third has to a fourth, and also a fifth has to the second the same ratio as a sixth to the fourth, then the sum of the first and fifth has to the second the same ratio as the sum of the third and sixth has to the fourth.

Let a first magnitude AB have to a second C the same ratio as a third DE has to a fourth F, and let also a fifth BG have to the second C the same ratio as a sixth EH has to the fourth F.

I say that the sum of the first and fifth, AG, has to the second C the same ratio as the sum of the third and sixth, DH, has to the fourth F.

Since BG is to C as EH is to F, inversely, C is to BG as F is to EH.	V.7.Cor
Then, since AB is to C as DE is to F, and C is to BG as F is to EH, therefore, *ex aequali*, AB is to BG as DE is to EH.	V.22

And, since the magnitudes are proportional taken separately, they are also proportional taken jointly, therefore AG is to GB as DH is to HE. V.18

But also BG is to C as EH is to F, therefore, *ex aequali*, AG is to C as DH is to F. V.22

Therefore, *if a first magnitude has to a second the same ratio as a third has to a fourth, and also a fifth has to the second the same ratio as a sixth to the fourth, then the sum of the first and fifth has to the second the same ratio as the sum of the third and sixth has to the fourth.*

Q.E.D.

Proposition 25

If four magnitudes are proportional, then the sum of the greatest and the least is greater than the sum of the remaining two.

Let the four magnitudes AB, CD, E, and F be proportional so that AB is to CD as E is to F, and let AB be the greatest of them and F the least.

I say that the sum of AB and F is greater than the sum of CD and E.

Make AG equal to E, and CH equal to F.	
Since AB is to CD as E is to F, and E equals AG, and F equals CH, therefore AB is to CD as AG is to CH.	V.7 V.11

And since the whole AB is to the whole CD as the part AG subtracted is to the part CH subtracted, therefore the remainder GB is also to the remainder HD as the whole AB is to the whole CD. V.19

But AB is greater than CD, therefore GB is also greater than HD. (V.14)

And, since *AG* equals *E*, and *CH* equals *F*, therefore the sum of *AG* and *F* equals the sum of *CH* and *E*.

And if, *GB* and *HD* being unequal, and *GB* greater, the sum of *AG* and *F* is added to *GB*, and the sum of *CH* and *E* is added to *HD*, it follows that the sum of *AB* and *F* is greater than the sum of *CD* and *E*.

Therefore, *if four magnitudes are proportional, then the sum of the greatest and the least is greater than the sum of the remaining two.*

Q.E.D.

Book VI Propositions

Proposition 14

In equal and equiangular parallelograms the sides about the equal angles are reciprocally proportional; and equiangular parallelograms in which the sides about the equal angles are reciprocally proportional are equal.

Let AB and BC be equal and equiangular parallelograms having the angles at B equal, and let DB and BE be placed in a straight line. Therefore FB and BG are also in a straight line.   I.14

I say that, in AB and BC, the sides about the equal angles are reciprocally proportional, that is to say, DB is to BE as BG is to BF.

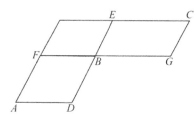

Complete the parallelogram FE.   I.31

Then since the parallelogram AB equals the parallelogram BC, and FE is another parallelogram, therefore AB is to FE as BC is to FE.   V.7

But AB is to FE as DB is to BE, and BC is to FE as BG is to BF. Therefore DB is to BE as BG is to BF.   VI.1, V.11

Therefore in the parallelograms AB and BC the sides about the equal angles are reciprocally proportional.

Next, let DB be to BE as BG is to BF.

I say that the parallelogram AB equals the parallelogram BC.

Since DB is to BE as BG is to BF, while DB is to BE as the parallelogram AB is to the parallelogram FE, and, BG is to BF as the parallelogram BC is to the parallelogram FE, therefore also AB is to FE as BC is to FE.   VI.1, V.11

Therefore the parallelogram AB equals the parallelogram BC.   V.9

Therefore, *in equal and equiangular parallelograms the sides about the equal angles are reciprocally proportional; and equiangular parallelograms in which the sides about the equal angles are reciprocally proportional are equal.*

Q.E.D.

Proposition 15

In equal triangles which have one angle equal to one angle the sides about the equal angles are reciprocally proportional; and those triangles which have one angle equal to one angle, and in which the sides about the equal angles are reciprocally proportional, are equal.

Let ABC and ADE be equal triangles having one angle equal to one angle, namely the angle BAC equal to the angle DAE.

I say that in the triangles ABC and ADE the sides about the equal angles are reciprocally proportional, that is to say, that CA is to AD as EA is to AB.

Place them so that CA is in a straight line with AD. Therefore EA is also in a straight line with AB.   I.14

Join BD.

Since then the triangle ABC equals the triangle ADE, and ABD is another triangle, therefore the triangle ABC is to the triangle ABD as the triangle ADE is to the triangle ABD.   V.7

But ABC is to ABD as AC is to AD, and ADE is to ABD as AE is to AB.   VI.1

Therefore also AC is to AD as AE is to AB.   V.11

Therefore in the triangles ABC and ADE the sides about the equal angles are reciprocally proportional.

Next, let the sides of the triangles ABC and ADE be reciprocally proportional, that is to say, let AE be to AB as CA is to AD.

I say that the triangle ABC equals the triangle ADE.

If BD is again joined, since AC is to AD as AE is to AB, while AC is to AD as the triangle ABC is to the triangle ABD, and AE is to AB as the triangle ADE is to the triangle ABD, therefore the triangle ABC is to the triangle ABD as the triangle ADE is to the triangle ABD.   VI.1
   V.11

Therefore each of the triangles ABC and ADE has the same ratio to ABD.

Therefore the triangle ABC equals the triangle ADE.   V.9

Therefore, *in equal triangles which have one angle equal to one angle the sides about the equal angles are reciprocally proportional; and those triangles which have one angle equal to one angle, and in which the sides about the equal angles are reciprocally proportional, are equal.*

Q.E.D.

Proposition 16

If four straight lines are proportional, then the rectangle contained by the extremes equals the rectangle contained by the means; and, if the rectangle contained by the extremes equals the rectangle contained by the means, then the four straight lines are proportional.

Let the four straight lines AB, CD, E, and F be proportional, so that AB is to CD as E is to F.
I say that the rectangle AB by F equals the rectangle CD by E.

Draw AG and CH from the points A and C at right angles to the straight lines AB and CD,   I.11
and make AG equal to F, and CH equal to E.   I.3

Complete the parallelograms BG and DH.   I.31

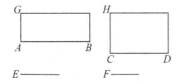

Then since AB is to CD as E is to F, while E equals CH, and F equals AG, therefore AB is to CD as CH is to AG.   V.7

Therefore in the parallelograms BG and DH the sides about the equal angles are reciprocally proportional.

But those equiangular parallelograms in which the sides about the equal angles are reciprocally proportional are equal, therefore the parallelogram *BG* equals the parallelogram *DH*. VI.14

And *BG* is the rectangle *AB* by *F*, for *AG* equals *F*, and *DH* is the rectangle *CD* by *E*, for *E* equals *CH*, therefore the rectangle *AB* by *F* equals the rectangle *CD* by *E*.

Next, let the rectangle *AB* by *F* be equal to the rectangle *CD* by *E*.
I say that the four straight lines are proportional, so that *AB* is to *CD* as *E* is to *F*.

With the same construction, since the rectangle *AB* by *F* equals the rectangle *CD* by *E*, and the rectangle *AB* by *F* is *BG*, for *AG* equals *F*, and the rectangle *CD* by *E* is *DH*, for *CH* equals *E*, therefore *BG* equals *DH*.

And they are equiangular. But in equal and equiangular parallelograms the sides about the equal angles are reciprocally proportional. VI.14

Therefore *AB* is to *CD* as *CH* is to *AG*. V.7

But *CH* equals *E*, and *AG* to *F*, therefore *AB* is to *CD* as *E* is to *F*.

Therefore, *if four straight lines are proportional, then the rectangle contained by the extremes equals the rectangle contained by the means; and, if the rectangle contained by the extremes equals the rectangle contained by the means, then the four straight lines are proportional.*

Q.E.D.

Proposition 17

If three straight lines are proportional, then the rectangle contained by the extremes equals the square on the mean; and, if the rectangle contained by the extremes equals the square on the mean, then the three straight lines are proportional.

Let the three straight lines *A* and *B* and *C* be proportional, so that *A* is to *B* as *B* is to *C*.
I say that the rectangle *A* by *C* equals the square on *B*.

Make *D* equal to *B*. I.3

Then, since *A* is to *B* as *B* is to *C*, and *B* equals *D*, therefore *A* is to *B* as *D* is to *C*. V.7
V.11

But, if four straight lines are proportional, then the rectangle contained by the extremes equals the rectangle contained by the means. VI.16

Therefore the rectangle *A* by *C* equals the rectangle *B* by *D*. But the rectangle *B* by *D* is the square on *B*, for *B* equals *D*, therefore the rectangle *A* by *C* equals the square on *B*.

Next, let the rectangle *A* by *C* equal the square on *B*.
I say that *A* is to *B* as *B* is to *C*.

With the same construction, since the rectangle *A* by *C* equals the square on *B*, while the square on *B* is the rectangle *B* by *D*, for *B* equals *D*, therefore the rectangle *A* by *C* equals the rectangle *B* by *D*.

But, if the rectangle contained by the extremes equals that contained by the means, then the four straight lines are proportional.   VI.16

Therefore *A* is to *B* as *D* is to *C*.

But *B* equals *D*, therefore *A* is to *B* as *B* is to *C*.

Therefore, *if three straight lines are proportional, then the rectangle contained by the extremes equals the square on the mean; and, if the rectangle contained by the extremes equals the square on the mean, then the three straight lines are proportional.*

Q.E.D.

Proposition 18

To describe a rectilinear figure similar and similarly situated to a given rectilinear figure on a given straight line.

Let *AB* be the given straight line and *CE* the given rectilinear figure.
It is required to describe on the straight line *AB* a rectilinear figure similar and similarly situated to the rectilinear figure *CE*.

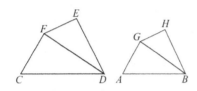

Join *DF*. Construct the angle *GAB* equal to the angle at *C*, and the angle *ABG* equal to the angle *CDF*, on the straight line *AB* at the points *A* and *B* on it.   I.23

Therefore the remaining angle *CFD* equals the angle *AGB*. Therefore the triangle *FCD* is equiangular with the triangle *GAB*.   I.32

Therefore, proportionally, *FD* is to *GB* as *FC* is to *GA*, and as *CD* is to *AB*.   VI.4, V.16

Again, construct the angle *BGH* equal to the angle *DFE*, and the angle *GBH* equal to the angle *FDE*, on the straight line *BG* and at the points *B* and *G* on it.   I.23

Therefore the remaining angle at *E* equals the remaining angle at *H*. Therefore the triangle *FDE* is equiangular with the triangle *GBH*. Therefore, proportionally, *FD* is to *GB* as *FE* is to *GH*, and as *ED* is to *HB*.   I.32, VI.4, V.16

But it was also proved that *FD* is to *GB* as *FC* is to *GA*, and as *CD* is to *AB*. Therefore *FC* is to *AG* as *CD* is to *AB*, and as *FE* is to *GH*, and further as *ED* is to *HB*.   V.11

And, since the angle *CFD* equals the angle *AGB*, and the angle *DFE* equals the angle *BGH*, therefore the whole angle *CFE* equals the whole angle *AGH*.
For the same reason the angle *CDE* also equals the angle *ABH*.
And the angle at *C* also equals the angle at *A*, and the angle at *E* equals the angle at *H*.

Therefore *AH* is equiangular with *CE*, and they have the sides about their equal angles proportional. Therefore the rectilinear figure *AH* is similar to the rectilinear figure *CE*.   VI.Def.1

Therefore the rectilinear figure *AH* has been described similar and similarly situated to the given rectilinear figure *CE* on the given straight line *AB*.

Q.E.F.

Proposition 19

Similar triangles are to one another in the duplicate ratio of the corresponding sides.

Let *ABC* and *DEF* be similar triangles having the angle at *B* equal to the angle at *E*, and such that *AB* is to *BC* as *DE* is to *EF*, so that *BC* corresponds to *EF*.   V.Def.11

I say that the triangle *ABC* has to the triangle *DEF* a ratio duplicate of that which *BC* has to *EF*.

Take a third proportional *BG* to *BC* and *EF* so that *BC* is to *EF* as *EF* is to *BG*, and join *AG*.   VI.11

Since *AB* is to *BC* as *DE* is to *EF*, therefore, alternately, *AB* is to *DE* as *BC* is to *EF*.   V.16

But *BC* is to *EF* as *EF* is to *BG*, therefore also *AB* is to *DE* as *EF* is to *BG*.   V.11

Therefore in the triangles *ABG* and *DEF* the sides about the equal angles are reciprocally proportional.

But those triangles which have one angle equal to one angle, and in which the sides about the equal angles are reciprocally proportional, are equal. Therefore the triangle *ABG* equals the triangle *DEF*.   VI.15

Now since *BC* is to *EF* as *EF* is to *BG*, and, if three straight lines are proportional, the first has to the third a ratio duplicate of that which it has to the second, therefore *BC* has to *BG* a ratio duplicate of that which *BC* has to *EF*.   V.Def.9

But *BC* is to *BG* as the triangle *ABC* is to the triangle *ABG*, therefore the triangle *ABC* also has to the triangle *ABG* a ratio duplicate of that which *BC* has to *EF*.   VI.1, V.11

But the triangle *ABG* equals the triangle *DEF*, therefore the triangle *ABC* also has to the triangle *DEF* a ratio duplicate of that which *BC* has to *EF*.   V.7

Therefore, *similar triangles are to one another in the duplicate ratio of the corresponding sides.*

Q.E.D.

Proposition 20

Similar polygons are divided into similar triangles, and into triangles equal in multitude and in the same ratio as the wholes, and the polygon has to the polygon a ratio duplicate of that which the corresponding side has to the corresponding side.

Let *ABCDE* and *FGHKL* be similar polygons, and let *AB* correspond to *FG*.
I say that the polygons *ABCDE* and *FGHKL* are divided into similar triangles, and into triangles equal in multitude and in the same ratio as the wholes, and the polygon *ABCDE* has to the polygon *FGHKL* a ratio duplicate of that which *AB* has to *FG*.
Join *BE*, *CE*, *GL*, and *HL*.

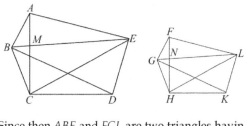

Now, since the polygon ABCDE is similar to the polygon FGHKL, therefore the angle BAE equals the angle GFL, and AB is to AE as GF is to FL.

VI.Def.1

Since then ABE and FGL are two triangles having one angle equal to one angle and the sides about the equal angles proportional, therefore the triangle ABE is equiangular with the triangle FGL, so that it is also similar, therefore the angle ABE equals the angle FGL.

VI.6
VI.4
VI.Def.1

But the whole angle ABC also equals the whole angle FGH because of the similarity of the polygons, therefore the remaining angle EBC equals the angle LGH.

And, since the triangles ABE and FGL are similar, BE is to AB as GL is to GF. Also, since the polygons are similar, AB is to BC as FG is to GH. Therefore, *ex aequali*, BE is to BC as GL is to GH, that is, the sides about the equal angles EBC and LGH are proportional. Therefore the triangle EBC is equiangular with the triangle LGH, so that the triangle EBC is also similar to the triangle LGH.

V.22
VI.6
VI.4
VI.Def.1

For the same reason the triangle ECD is also similar to the triangle LHK.
Therefore the similar polygons ABCDE and FGHKL have been divided into similar triangles, and into triangles equal in multitude.
I say that they are also in the same ratio as the wholes, that is, in such manner that the triangles are proportional, and ABE, EBC, and ECD are antecedents, while FGL,

LGH, and LHK are their consequents, and that the polygon ABCDE has to the polygon FGHKL a ratio duplicate of that which the corresponding side has to the corresponding side, that is AB to FG.
Join AC and FH.

Then since the polygons are similar, the angle ABC equals the angle FGH, and AB is to BC as FG is to GH, the triangle ABC is equiangular with the triangle FGH, therefore the angle BAC equals the angle GFH, and the angle BCA to the angle GHF.

VI.6

And, since the angle BAM equals the angle GFN, and the angle ABM also equals the angle FGN, therefore the remaining angle AMB also equals the remaining angle FNG. Therefore the triangle ABM is equiangular with the triangle FGN.

I.32

Similarly we can prove that the triangle BMC is also equiangular with the triangle GNH.

Therefore, proportionally, AM is to MB as FN is to NG, and BM is to MC as GN is to NH. So that, in addition, *ex aequali*, AM is to MC as FN is to NH.

V.22

But AM is to MC as the triangle ABM is to MBC, and as AME is to EMC, for they are to one another as their bases.

VI.1

Therefore also one of the antecedents is to one of the consequents as are all the antecedents to all the consequents, therefore the triangle AMB is to BMC as ABE is to CBE.

V.12

But *AMB* is to *BMC* as *AM* is to *MC*, therefore *AM* is to *MC* as the triangle *ABE* is to the triangle *EBC*. V.11

For the same reason also *FN* is to *NH* as the triangle *FGL* is to the triangle *GLH*.

And *AM* is to *MC* as *FN* is to *NH*, therefore the triangle *ABE* is to the triangle *BEC* as the triangle *FGL* is to the triangle *GLH*, and, alternately, the triangle *ABE* is to the triangle *FGL* as the triangle *BEC* is to the triangle *GLH*. V.11
V.16

Similarly we can prove, if *BD* and *GK* are joined, that the triangle *BEC* is to the triangle *LGH* as the triangle *ECD* is to the triangle *LHK*.

And since the triangle *ABE* is to the triangle *FGL* as *EBC* is to *LGH*, and further as *ECD* is to *LHK*, therefore also one of the antecedents is to one of the consequents as the sum of the antecedents to the sum of the consequents. Therefore the triangle *ABE* is to the triangle *FGL* as the polygon *ABCDE* is to the polygon *FGHKL*. V.12

But the triangle *ABE* has to the triangle *FGL* a ratio duplicate of that which the corresponding side *AB* has to the corresponding side *FG*, for similar triangles are in the duplicate ratio of the corresponding sides. VI.19

Therefore the polygon *ABCDE* also has to the polygon *FGHKL* a ratio duplicate of that which the corresponding side *AB* has to the corresponding side *FG*. V.11

Therefore, *similar polygons are divided into similar triangles, and into triangles equal in multitude and in the same ratio as the wholes, and the polygon has to the polygon a ratio duplicate of that which the corresponding side has to the corresponding side.*

Q.E.D.

Proposition 21

Figures which are similar to the same rectilinear figure are also similar to one another.

Let each of the rectilinear figures *A* and *B* be similar to *C*.
I say that A is also similar to B.

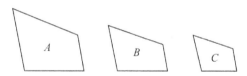

Since *A* is similar to *C*, it is equiangular with it and has the sides about the equal angles proportional. VI.Def.1

Again, since *B* is similar to *C*, it is equiangular with it and has the sides about the equal angles proportional.

Therefore each of the figures *A* and *B* is equiangular with *C* and with *C* has the sides about the equal angles proportional, therefore *A* is similar to *B*. V.11

Therefore, *figures which are similar to the same rectilinear figure are also similar to one another.*

Q.E.D.

Proposition 22

If four straight lines are proportional, then the rectilinear figures similar and similarly described upon them are also proportional; and, if the rectilinear figures similar and similarly described upon them are proportional, then the straight lines are themselves also proportional.

Let the four straight lines AB, CD, EF, and GH be proportional, so that AB is to CD as EF is to GH. Let the similar and similarly situated rectilinear figures KAB and LCD be described on AB and CD, and the similar and similarly situated rectilinear figures MF and NH be described on EF and GH.

I say that KAB is to LCD as MF is to NH.

Take a third proportional O to AB and CD, and a third proportional P to EF and GH.   VI.11

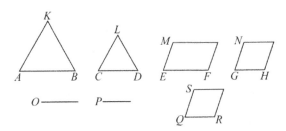

Then since AB is to CD as EF is to GH, therefore CD is to O as GH is to P. Therefore, ex aequali, AB is to O as EF is to P.	V.11 V.22
But AB is to O as KAB is to LCD, and EF is to P as MF is to NH, therefore KAB is to LCD also as MF is to NH.	VI.19,Cor V.11

Next, let KAB be to LCD as MF is to NH.
I say also that AB is to CD as EF is to GH.

For, if EF is not to GH as AB is to CD, let EF be to QR as AB is to CD. Describe the rectilinear figure SR similar and similarly situated to either of the two MF or NH on QR.	VI.12 VI.18

Since then AB is to CD as EF is to QR, and there have been described on AB and CD the similar and similarly situated figures KAB and LCD, and on EF and QR the similar and similarly situated figures MF and SR, therefore KAB is to LCD as MF is to SR.

But also, by hypothesis, KAB is to LCD as MF is to NH, therefore also MF is to SR as MF is to NH.   V.11

Therefore MF has the same ratio to each of the figures NH and SR, therefore NH equals SR.   V.9

But it is also similar and similarly situated to it, therefore GH equals QR.
And, since AB is to CD as EF is to QR, while QR equals GH, therefore AB is to CD as EF is to GH.

Therefore, *if four straight lines are proportional, then the rectilinear figures similar and similarly described upon them are also proportional; and, if the rectilinear figures similar and similarly described upon them are proportional, then the straight lines are themselves also proportional.*

Q.E.D.

Proposition 23

Equiangular parallelograms have to one another the ratio compounded of the ratios of their sides.

Let AC and CF be equiangular parallelograms having the angle BCD equal to the angle ECG.

I say that the parallelogram AC has to the parallelogram CF the ratio compounded of the ratios of the sides.

Let them be placed so that BC is in a straight line with CG. Then DC is also in a straight line with CE. I.14

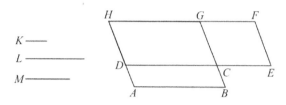

Complete the parallelogram DG. Set out a straight line K, and make it so that BC is to CG as K is to L, and DC is to CE as L is to M. I.31 VI.12

Then the ratios of K to L and of L to M are the same as the ratios of the sides, namely of BC to CG and of DC to CE.

But the ratio of K to M is compounded of the ratio of K to L and of that of L to M, so that K has also to M the ratio compounded of the ratios of the sides.

Now since BC is to CG as the parallelogram AC is to the parallelogram CH, and BC is to CG as K is to L, therefore K is to L as AC is to CH. VI.1 V.11

Again, since DC is to CE as the parallelogram CH is to CF, and DC is to CE as L is to M, therefore L is to M as the parallelogram CH is to the parallelogram CF. VI.1 V.11

Since then it was proved that K is to L as the parallelogram AC is to the parallelogram CH, and L is to M as the parallelogram CH is to the parallelogram CF, therefore, *ex aequali* K is to M as AC is to the parallelogram CF. V.22

But K has to M the ratio compounded of the ratios of the sides, therefore AC also has to CF the ratio compounded of the ratios of the sides.

Therefore, *equiangular parallelograms have to one another the ratio compounded of the ratios of their sides.*

Q.E.D.

Proposition 24

In any parallelogram the parallelograms about the diameter are similar both to the whole and to one another.

Let ABCD be a parallelogram, and AC its diameter, and let EG and HK be parallelograms about AC.

I say that each of the parallelograms EG and HK is similar both to the whole ABCD and to the other.

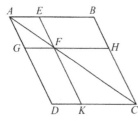

For, since EF is parallel to a side BC of the triangle ABC, proportionally, BE is to EA as CF is to FA. — VI.2

Again, since FG is parallel to a side CD of the triangle ACD, proportionally, CF is to FA as DG is to GA. — VI.2

But it was proved that CF is to FA as BE is to EA, therefore BE is to EA as DG is to GA. Therefore, taken jointly, BA is to AE as DA is to AG, and, alternately, BA is to AD as EA is to AG. — V.18, V.16

Therefore in the parallelograms ABCD and EG, the sides about the common angle BAD are proportional.

And, since GF is parallel to DC, the angle AFG equals the angle ACD, and the angle DAC is common to the two triangles ADC and AGF, therefore the triangle ADC is equiangular with the triangle AGF. — I.29

For the same reason the triangle ACB is also equiangular with the triangle AFE, and the whole parallelogram ABCD is equiangular with the parallelogram EG.

Therefore, proportionally, AD is to DC as AG is to GF, DC is to CA as GF is to FA, AC is to CB as AF is to FE, and CB is to BA as FE is to EA.

And, since it was proved that DC is to CA as GF is to FA, and AC is to CB as AF is to FE, therefore, *ex aequali*, DC is to CB as GF is to FE. — V.22

Therefore in the parallelograms ABCD and EG the sides about the equal angles are proportional. Therefore the parallelogram ABCD is similar to the parallelogram EG. — VI.Def. 1

For the same reason the parallelogram ABCD is also similar to the parallelogram KH. Therefore each of the parallelograms EG and HK is similar to ABCD.

But figures similar to the same rectilinear figure are also similar to one another, therefore the parallelogram EG is also similar to the parallelogram HK. — VI.21

Therefore, *in any parallelogram the parallelograms about the diameter are similar both to the whole and to one another*.

Q.E.D.

Proposition 25

To construct a figure similar to one given rectilinear figure and equal to another.

Let ABC be the given rectilinear figure to which the figure to be constructed must be similar, and D that to which it must be equal.

It is required to construct one figure similar to ABC and equal to D.

Let there be applied to BC the parallelogram BE equal to the triangle ABC, and to CE the parallelogram CM equal to D in the angle FCE which equals the angle CBL. — I.44, I.45

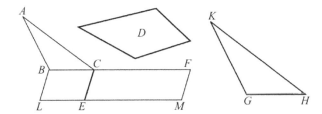

Then *BC* is in a straight line with *CF*, and *LE* with *EM*.

Take a mean proportional *GH* to *BC* and *CF*, and describe *KGH* similar and similarly situated to *ABC* on *GH*. VI.13 VI.18

Then, since *BC* is to *GH* as *GH* is to *CF*, and, if three straight lines are proportional, then the first is to the third as the figure on the first is to the similar and similarly situated figure described on the second, therefore *BC* is to *CF* as the triangle *ABC* is to the triangle *KGH*. V.19,Cor

But *BC* is to *CF* as the parallelogram *BE* is to the parallelogram *EF*. VI.1

Therefore also the triangle *ABC* is to the triangle *KGH* as the parallelogram *BE* is to the parallelogram *EF*. Therefore, alternately, the triangle *ABC* is to the parallelogram *BE* as the triangle *KGH* is to the parallelogram *EF*. V.11 V.16

But the triangle *ABC* equals the parallelogram *BE*, therefore the triangle *KGH* also equals the parallelogram *EF*. And the parallelogram *EF* equals *D*, therefore *KGH* also equals *D*. (V.14)

And *KGH* is also similar to *ABC*. Therefore this figure *KGH* has been constructed similar to the given rectilinear figure *ABC* and equal to the other given figure *D*.

Q.E.F.

Proposition 26

If from a parallelogram there is taken away a parallelogram similar and similarly situated to the whole and having a common angle with it, then it is about the same diameter with the whole.

From the parallelogram *ABCD* let there be taken away the parallelogram *AF* similar and similarly situated to *ABCD*, and having the angle *DAB* common with it.
I say that *ABCD* is about the same diameter with *AF*.

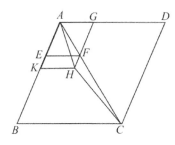

For suppose it is not, but, if possible, let *AHC* be the diameter. Produce *GF* and carry it through to *H*. Draw *HK* through *H* parallel to either of the straight lines *AD* or *BC*. I.31

Since, then, *ABCD* is about the same diameter with *KG*, therefore *DA* is to *AB* as *GA* is to *AK*. VI.24

But also, since ABCD and EG are similar, therefore DA is to AB as GA is to AE.　　VI.Def.1
Therefore GA is to AK as GA is to AE.　　V.11

Therefore GA has the same ratio to each of the straight lines AK and AE.

Therefore AE equals AK the less equals the greater, which is impossible.　　V.9

Therefore ABCD cannot fail to be about the same diameter with AF. Therefore the parallelogram ABCD is about the same diameter with the parallelogram AF.

Therefore, *if from a parallelogram there is taken away a parallelogram similar and similarly situated to the whole and having a common angle with it, then it is about the same diameter with the whole.*

Q.E.D.

Proposition 27

Of all the parallelograms applied to the same straight line falling short by parallelogrammic figures similar and similarly situated to that described on the half of the straight line, that parallelogram is greatest which is applied to the half of the straight line and is similar to the difference.

Let AB be a straight line and let it be bisected at C. Let there be applied to the straight line AB the parallelogram AD falling short by the parallelogrammic figure DB described on the half of AB, that is, CB.

I say that, of all the parallelograms applied to AB falling short by parallelogrammic figures similar and similarly situated to DB, AD is greatest.

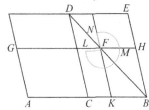

Let there be applied to the straight line AB the parallelogram AF falling short by the parallelogrammic figure FB similar and similarly situated to DB.

I say that AD is greater than AF.

Since the parallelogram DB is similar to the parallelogram FB, therefore they are about the same diameter.　　VI.26

Draw their diameter DB, and describe the figure.

Then, since CF equals FE, and FB is common, therefore the whole CH equals the whole KE.　　I.43

But CH equals CG, since AC also equals CB.　　I.36

Therefore CG also equals KE.

Add CF to each. Therefore the whole AF equals the gnomon LMN, so that the parallelogram DB, that is, AD, is greater than the parallelogram AF.

Therefore, *of all the parallelograms applied to the same straight line falling short by parallelogrammic figures similar and similarly situated to that described on the half of the straight line, that parallelogram is greatest which is applied to the half of the straight line and is similar to the difference.*

Q.E.D.

Proposition 28

To apply a parallelogram equal to a given rectilinear figure to a given straight line but falling short by a parallelogram similar to a given one; thus the given rectilinear figure must not be greater than the parallelogram described on the half of the straight line and similar to the given parallelogram.

VI.27

Let C be the given rectilinear figure, AB the given straight line, and D the given parallelogram, and let C not be greater than the parallelogram described on the half of AB similar to the given parallelogram D.

It is required to apply a parallelogram equal to the given rectilinear figure C to the given straight line AB but falling short by a parallelogram similar to D.

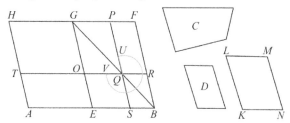

Bisect AB at the point E. Describe $EBFG$ similar and similarly situated to D on EB, and complete the parallelogram AG.

I.9
VI.18

If then AG equals C, that which was proposed is done, for the parallelogram AG equal to the given rectilinear figure C has been applied to the given straight line AB but falling short by a parallelogram GB similar to D.

But, if not, let HE be greater than C.

Now HE equals GB, therefore GB is also greater than C.

Construct $KLMN$ equal to GB minus C and similar and similarly situated to D.

VI.25

But D is similar to GB, therefore KM is also similar to GB.

VI.21

Let, then, KL correspond to GE, and LM to GF.

Now, since GB equals C and KM, therefore GB is greater than KM, therefore also GE is greater than KL, and GF than LM.

Make GO equal to KL, and GP equal to LM, and let the parallelogram $OGPQ$ be completed, therefore it is equal and similar to KM.

Therefore GQ is also similar to GB, therefore GQ is about the same diameter with GB.

VI.21
VI.26

Let GQB be their diameter, and describe the figure. Then, since BG equals C and KM, and in them GQ equals KM, therefore the remainder, the gnomon UWV, equals the remainder C. And, since PR equals OS, add QB to each, therefore the whole PB equals the whole OB.

But OB equals TE, since the side AE also equals the side EB, therefore TE also equals PB.

I.36

Add *OS* to each. Therefore the whole *TS* equals the whole, the gnomon *VWU*.
But the gnomon *VWU* was proved equal to *C*, therefore *TS* also equals *C*.
Therefore there the parallelogram *ST* equal to the given rectilinear figure *C* has been applied to the given straight line *AB* but falling short by a parallelogram *QB* similar to *D*.

Q.E.F.

Proposition 29

To apply a parallelogram equal to a given rectilinear figure to a given straight line but exceeding it by a parallelogram similar to a given one.

Let *C* be the given rectilinear figure, *AB* be the given straight line, and *D* the parallelogram to which the excess is required to be similar.

It is required to apply a parallelogram equal to the the rectilinear figure *C* to the straight line *AB* but exceeding it by a parallelogram similar to *D*.

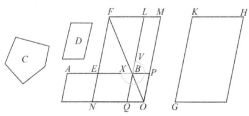

Bisect *AB* at *E*. Describe the parallelogram *BF* on *EB* similar and similarly situated to *D*, and construct *GH* equal to the sum of *BF* and *C* and similar and similarly situated to *D*.	VI.25

Let *KH* correspond to *FL* and *KG* to *FE*.

Now, since *GH* is greater than *FB*, therefore *KH* is also greater than *FL*, and *KG* greater than *FE*.

Produce *FL* and *FE*. Make *FLM* equal to *KH*, and *FEN* equal to *KG*. Complete *MN*.
Then *MN* is both equal and similar to *GH*.

But *GH* is similar to *EL*, therefore *MN* is also similar to *EL*, therefore *EL* is about the same diameter with *MN*.	VI.21 VI.26

Draw their diameter *FO*, and describe the figure.

Since *GH* equals the sum of *EL* and *C*, while *GH* equals *MN*, therefore *MN* also equals the sum of *EL* and *C*.

Subtract *EL* from each. Therefore the remainder, the gnomon *XWV*, equals *C*.

Now, since *AE* equals *EB*, therefore *AN* equals *NB*, that is, *LP*.	I.36 I.43

Add *EO* to each. Therefore the whole *AO* equals the gnomon *VWX*.
But the gnomon *VWX* equals *C*, therefore *AO* also equals *C*.

Therefore the parallelogram *AO* equal to the given rectilinear figure *C* has been applied to the given straight line *AB* but exceeding it by a parallelogram *QP* similar to *D*, since *PQ* is also similar to *EL*.	VI.24

Q.E.F.

Proposition 30

To cut a given finite straight line in extreme and mean ratio.

Let *AB* be the given finite straight line.
It is required to cut *AB* in extreme and mean ratio.

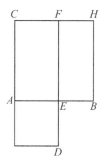

Describe the square *BC* on *AB*. Apply the parallelogram *CD* to *AC* equal to the sum of *BC* and the figure *AD* similar to *BC*.	I.46 VI.29
Now *BC* is a square, therefore *AD* is also a square. And, since *BC* equals *CD*, subtract *CE* from each, therefore the remainder *BF* equals the remainder *AD*.	
But it is also equiangular with it, therefore in *BF* and *AD* the sides about the equal angles are reciprocally proportional. Therefore *FE* is to *ED* as *AE* is to *EB*.	VI.14
But *FE* equals *AB*, and *ED* equals *AE*.	

Therefore *AB* is to *AE* as *AE* is to *EB*. V.7
And *AB* is greater than *AE*, therefore *AE* is also greater than *EB*.
Therefore the straight line *AB* has been cut in extreme and mean ratio at *E*, and the greater segment of it is *AE*. VI.Def.3

Q.E.F.

Proposition 31

In right-angled triangles the figure on the side opposite the right angle equals the sum of the similar and similarly described figures on the sides containing the right angle.

Let *ABC* be a right-angled triangle having the angle *BAC* right.
I say that the figure on *BC* equals the sum of the similar and similarly described figures on *BA* and *AC*.
Draw the perpendicular *AD*. I.12

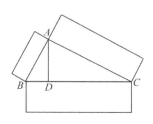

Then, since in the right-angled triangle *ABC*, *AD* has been drawn from the right angle at *A* perpendicular to the base *BC*, therefore the triangles *DBA* and *DAC* adjoining the perpendicular are similar both to the whole *ABC* and to one another.	VI.8
And, since *ABC* is similar to *DBA*, therefore *BC* is to *BA* as *BA* is to *BD*.	VI.Def.1
And, since three straight lines are proportional, the first is to the third as the figure on the first is to the similar and similarly described figure on the second.	VI.19,Cor

Therefore BC is to BD as the figure on BC is to the similar and similarly described figure on BA.

For the same reason also, BC is to CD as the figure on BC is to that on CA, so that, in addition, BC is to the sum of BD and DC as the figure on BC is to the sum of the similar and similarly described figures on BA and AC. V.24

But BC equals the sum of BD and DC, therefore the figure on BC equals the sum of the similar and similarly described figures on BA and AC.

Therefore, *in right-angled triangles the figure on the side opposite the right angle equals the sum of the similar and similarly described figures on the sides containing the right angle.*

Q.E.D.

Proposition 32

If two triangles having two sides proportional to two sides are placed together at one angle so that their corresponding sides are also parallel, then the remaining sides of the triangles are in a straight line.

Let ABC and DCE be two triangles having the two sides AB and AC proportional to the two sides DC and DE, so that AB is to AC as DC is to DE, and AB parallel to DC, and AC parallel to DE.

I say that BC is in a straight line with CE.

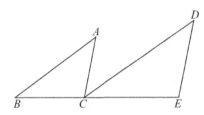

Since AB is parallel to DC, and the straight line AC falls upon them, therefore the alternate angles BAC and ACD equal one another. I.29

For the same reason the angle CDE also equals the angle ACD, so that the angle BAC equals the angle CDE.

And, since ABC and DCE are two triangles having one angle, the angle at A, equal to one angle, the angle at D, and the sides about the equal angles proportional, so that AB is to AC as DC is to DE, therefore the triangle ABC is equiangular with the triangle DCE. Therefore the angle ABC equals the angle DCE. VI.6

But the angle ACD was also proved equal to the angle BAC, therefore the whole angle ACE equals the sum of the two angles ABC and BAC.

Add the angle ACB to each. Therefore the sum of the angles ACE and ACB equals the sum of the angles BAC, ACB, and CBA.

But the sum of the angles BAC, ABC, and ACB equals two right angles, therefore the sum of the angles ACE and ACB also equals two right angles. I.32

Therefore with a straight line *AC*, and at the point *C* on it, the two straight lines *BC* and
CE not lying on the same side make the sum of the adjacent angles *ACE* and *ACB* equal to   I.14
two right angles. Therefore *BC* is in a straight line with *CE*.

Therefore, *if two triangles having two sides proportional to two sides are placed together at one angle so
that their corresponding sides are also parallel, then the remaining sides of the triangles are in a straight
line.*

Q.E.D.

Proposition 33

*Angles in equal circles have the same ratio as the circumferences on which they stand whether they stand
at the centers or at the circumferences.*

Let *ABC* and *DEF* be equal circles, and let the angles *BGC* and *EHF* be angles at their centers *G*
and *H*, and the angles *BAC* and *EDF* angles at the circumferences.

I say that the circumference *BC* is to the circumference *EF* as the angle *BGC* is to the angle *EHF*,
and as the angle *BAC* is to the angle *EDF*.

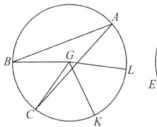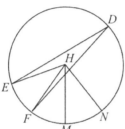

Make any number of consecutive circumferences *CK* and *KL* equal to the circumference *BC*, and any number of consecutive circumferences *FM, MN* equal to the circumference *EF*, and join *GK* and *GL* and *HM* and *HN*.

Then, since the circumferences *BC, CK,* and *KL* equal one another, the angles *BGC,
CGK,* and *KGL* also equal one another. Therefore, whatever multiple the   III.27
circumference *BL* is of *BC*, the angle *BGL* is also that multiple of the angle *BGC*.

For the same reason, whatever multiple the circumference *NE* is of *EF*, the angle
NHE is also that multiple of the angle *EHF*.

If the circumference *BL* equals the circumference *EN*, then the angle *BGL* also equals
the angle *EHN*; if the circumference *BL* is greater than the circumference *EN*, then the   III.27
angle *BGL* is also greater than the angle *EHN*; and, if less, less.

There being then four magnitudes, two circumferences *BC* and *EF*, and two angles
BGC and *EHF*, there have been taken, of the circumference *BC* and the angle *BGC*
equimultiples, namely the circumference *BL* and the angle *BGL*, and of the
circumference *EF* and the angle *EHF* equimultiples, namely the circumference *EN*
and the angle *EHN*.

And it has been proved that, if the circumference *BL* is in excess of the circumference
EN, the angle *BGL* is also in excess of the angle *EHN*; if equal, equal; and if less, less.

Therefore the circumference *BC* is to *EF* as the angle *BGC* is to the angle *EHF*.   V.Def.5

But the angle *BGC* is to the angle *EHF* as the angle *BAC* is to the angle *EDF*, for they are doubles respectively.   V.15 / III.20

Therefore also the circumference *BC* is to the circumference *EF* as the angle *BGC* is to the angle *EHF*, and the angle *BAC* to the angle *EDF*.

Therefore, *angles in equal circles have the same ratio as the circumferences on which they stand whether they stand at the centers or at the circumferences.*

Q.E.D.

Appendix B

Logical Beginnings

Two Types of Reasoning

There are, in the broadest sense, two ways that we reason. Sometimes we draw a general conclusion from observations we have made. For example, we believe that the sun will rise tomorrow morning based on the fact that we know it has risen at least as long as men have been on the earth. And sometimes we believe a particular statement to be true because we first accept certain other statements as being true and then we think our way to the conclusion that the original statement is true. For example, if we believe the following two statements to be true

> All cats have four legs
> Tabby is a cat

then we also believe the statement

> Tabby has four legs.

Furthermore, our belief of the last statement is based on the fact that we believe the first two statements.

These two kinds of reasoning have two different names. We call the first kind **induction**, or inductive reasoning, and the second kind **deduction**, or deductive reasoning. We reason inductively when we draw conclusions from observations. We reason deductively when we use rules of logic. Mathematical reasoning is deductive reasoning and it is this kind of reasoning you are going to be introduced to now. You are going to be introduced to the very beginning of the study of logic. This will enable you to better analyze Euclid's propositions and better understand the reasoning (or logic) used to prove them.

Arguments

The concept of an argument is one most people are pretty familiar with. We tend to think of arguments as disagreements, or quarrels. In logical terms, though, "argument" takes on a different meaning. An **argument** in logic is simply a collection of statements, one of which can be singled out and called the **conclusion**, and the others of which are meant to support the conclusion. This definition holds whether we are reasoning inductively or deductively. So when we argue (inductively) that the sun will rise tomorrow because it has risen at least as long as men have been on the earth, our argument consists of two statements,

> The sun will rise tomorrow.
> The sun has risen at least as long as men have been on the earth.

In this case, the first of these statements is the conclusion of the argument. It is what we believe and, perhaps, are trying to convince someone else to believe. The second statement is meant to support the conclusion, that is, it is meant to give a person reason to think the conclusion is true.

Again look at the deductive argument stated above:

> All cats have four legs
> Tabby is a cat
> Tabby has four legs

You can see that the conclusion is the last statement, "Tabby has four legs," and you can see that this conclusion is supported by the two statements that precede it.

Conclusions

In order to analyze any argument, inductive or deductive, mathematical or not, the first and most important thing you need to be able to do is identify the conclusion of the argument. Most arguments are not as simple as the above two examples. Many are made up of sub-arguments whose threads can be difficult to follow. If you are able to identify the conclusion of an argument, then you can try to decide whether or not the author has adequately supported that conclusion with his other statements. But if you are not clear on which statement is the conclusion you cannot possibly decide whether it has been adequately supported and therefore, whether or not you have a good argument.

See if you can identify the conclusions in the following statements. If you find you are having trouble, reword each statement so that it is in the form "If…then…"

1. If $x = y$, then $x + z = y + z$.

2. Given two constants a and b, a + b cannot be a variable.

3. All even numbers are divisible by two.

4. $T = kVP$ implies $V = T/kP$.

Having identified the conclusions of these statements, you have noticed, of course, that there is a condition that makes the conclusion true. This condition is called the **hypothesis**. If there is more than one condition they are called **hypotheses**. But note that even though there may be several hypotheses in an argument, there can never be more than one conclusion.

Because the truth of the conclusion of each of the above statements is dependent on the truth of the corresponding hypothesis, these statements are called **conditional statements**. That is, there is a condition that must be satisfied before the conclusion can be known to be true. A conditional statement, then, can always be put in the form of two statements which are connected by "if" and "then."

The four statements on the previous page demonstrate four very common ways of making conditional statements. The ways you will see most often in this class are like statements 1 and 2 and variations of statement 3 ("Isosceles triangles have two equal angles" would be an example of a variation. Such statements are called **categorical** in nature.)

Now we will take some statements from Euclid and ask you to find the conclusions and put the statements in "If...then" form (if they aren't already in that form). Remember, you do not necessarily need to know the meanings of all the words in order to be able to find the conclusion and change the form of the statement. Here they are:

5. In isosceles triangles the angles at the base equal one another.

6. If two triangles have two angles equal to two angles respectively, and one side equal to one side, namely, either the side adjoining the equal angles, or that opposite one of the equal angles, then the remaining sides equal the remaining sides and the remaining angle equals the remaining angle.

7. To construct a square on a given finite straight line.

8. In any triangle the side opposite the greater angle is greater. (Be careful.)

There are two other forms of expressing conditional statements that you may see. Both of them actually split the statement into its hypothesis (or hypotheses) and conclusion and thus make it very simple to distinguish between hypotheses and conclusions.

9. Given: A triangle with two equal angles
To Prove: The sides opposite the equal angles are equal.

10. Hypothesis: AB is a finite straight line.
Conclusion: It is possible to construct a triangle on line AB.

You have now seen ten examples of conditional statements, many of which a student of mathematics might like to see proved. You have learned that there are different ways of expressing conditional statements. You have learned that the most common way of writing conditional statements is in "If...then..." form, and you have had some practice in changing statements into this form. The propositions you will encounter while studying the *Elements* are almost always (except for constructions) expressed either as "If...then..." statements or in what has been referred to as categorical form.

You should be certain you understand how to convert any conditional statement into "If...then..." form, because if you cannot do that, then there is considerable doubt as to whether you have accurately identified the conclusion and the hypothesis of the statement. To get additional practice in this skill, you can take any of the propositions of Euclid (other than constructions) and write them as "If...then..." statements.

Direct and indirect proofs

As you study the *Elements* you will encounter two different types of proofs. They are called **direct** proofs and **indirect** proofs. Direct proofs follow a line of reasoning based on definitions, postulates and common notions, making no assumptions along the way other than those made in the hypotheses of the statement to be proved. We will illustrate with an example of a very simple direct proof from elementary algebra.

If $x + 5 = 7$, then $x = 2$

A short direct proof might look something like the following. (A longer proof would note the existence of an additive identity and additive inverses as well as the associativity of addition. If you are familiar with these concepts, you may write out a more detailed proof.)

Given: $x + 5 = 7$
Prove: $x = 2$

Statement	Reason
1. $x + 5 = 7$	given
2. $x + 5 - 5 = 7 - 5$	if equals are subtracted from equals the results are equal
3. $x = 2$	simplify
	Q.E.D.

Unfortunately, some statements are not very easy (and some, in fact, impossible) to prove directly. Sometimes we have to use a technique called an indirect proof to arrive at our conclusion. When we prove something indirectly we assume the conclusion to be false, and with that assumption, using nothing but our definitions, common notions and reasoning, we arrive at something we know to be false. Since the only possible error we made was our assumption of a false conclusion, the conclusion cannot be false, and therefore must be true. An example or two will go a long way toward explaining the situation.

The first example we are going to give is one from ordinary arguments (as in disagreements) in which one person wants to convince another of something. Suppose your brother gets all A's at school and everyone knows it. The school bully walks in and you and he have the following verbal exchange.

> Bully: Man, your brother is really stupid.
> You: Oh, yeah? If he were stupid he wouldn't get straight A's!

You have just provided an indirect proof that Mr. Bully is wrong. What you want to prove is that your brother is smart. But you have not said, in a direct fashion, "My brother gets straight A's. Therefore, he is smart" (which sounds a bit weak, somehow, in a heated debate).

What you have said, in effect, is, "Suppose he were stupid. Then he wouldn't get straight A's." You, your friends, and Mr. Bully all know your brother does get straight A's. By assuming the truth of Mr. Bully's statement you have arrived at a contradiction of something everyone knows to be true, namely, that your brother gets straight A's.

Now the truth of Mr. Bully's statement is also the negation of the conclusion of your own argument. So you have supposed the negation of your own conclusion to be true, and then arrived at a contradiction. In informal argument your statements are sufficient. From the standpoint of formal argument, however, you need to say a little more. Your full argument, with the "understood" statements in parentheses, is the following.

> Suppose my brother were stupid.
> Then he wouldn't get straight A's.
> (But he does get straight A's.
> Therefore, he is not stupid.)

In an informal argument, oftentimes statements that are understood by all parties are omitted, with everyone understanding that without them the argument is not complete.

Let's go over carefully one more time what you did. You set out to prove your brother was not stupid. You assumed he was, then deduced that he would not, under those circumstances, get straight A's. But this is a contradiction to the fact that he does get straight A's. Having deduced from the assumption of his being stupid a fact that contradicts something everyone knows is true, you conclude that the original assumption of stupidity must be wrong.

You have now seen the basic plan of an indirect argument. It is something we use very naturally in informal speech, and which takes far longer to explain than to understand in that context. But sometimes in a formal context it can cause confusion. Let's go back to our mathematical example and write an indirect proof for it.

Given: $x+5=7$
Prove: $x=2$

Proof:

	Statement	Reason
1.	$x+5=7$	given
2.	Suppose $x \neq 2$	assumption we hope will lead to a contradiction
3.	$x>2$ or $x<2$	for real numbers a and b, either a = b, a > b, or a < b
4.	Let $x > 2$	pick either case
5.	$x+5 > 2+5$	if a number (5) is added to both sides of an inequality, the inequality is preserved.
7.	$x+5 > 7$	simplify

But $x+5>7$ contradicts the hypothesis $x+5=7$. Since our reasoning is correct, and we did not make any assumptions other than $x > 2$, we must conclude that this assumption is wrong.
In a similar fashion we can show that x cannot be less than 2.
Therefore, since x can be neither greater nor less than 2, $x = 2$.
Q.E.D.

Notice how much more complicated this proof is than the direct one. You can see that it is important to choose carefully how to go about proving something. For most propositions a direct proof is preferable, but sometimes an indirect one is unavoidable.

Let's go over one last time the overall scheme of an indirect proof. You assume the conclusion you believe to be true is actually false. Then by reasoning things through you arrive at a contradiction of something you know to be true (often, but by no means always, the hypothesis of your statement). Since the only possible error you made was your assumption that your conclusion is false, it must be at that point that you are wrong. Therefore, your conclusion is true.

Contrapositive and Converse

In the previous section we proved indirectly that if $x+5=7$, then $x = 2$. We assumed the conclusion, $x = 2$, was false and then we deduced that $x+5 \neq 7$ which is the negation of our hypothesis. So the contradiction in this indirect proof was the negation of the hypothesis.[1] In other words, in this particular case, in order to prove that the hypothesis implied the conclusion, what we ended up proving was that the negation of the conclusion implied the negation of the hypothesis.

This new statement, created out of a conditional statement, is encountered so often that it has a special name. It is called the **contrapositive** of the original statement. We state the following definition:

> The **contrapositive** of a conditional statement is the new statement obtained by negating the conclusion of the original statement and making it the hypothesis of the new statement, and negating the hypothesis of the original statement and making it the conclusion of the new statement.

As is so often the case, a few examples will make things much clearer.

Consider the statement "If Tabby is a cat, then Tabby has four legs." Using the above definition, the contrapositive of this statement is, "If Tabby does not have four legs, then Tabby is not a cat." Here are a few more statements for you to consider.

Statement: If you like pizza, then you must love Uno's.
Contrapositive: If you don't love Uno's, then you don't like pizza.

Statement: If a triangle has two equal angles, then it has two equal sides.
Contrapositive: If a triangle does not have two equal sides, then it does not have two equal angles.

---

[1] This is not always the case. Many times the contradiction will be of some known mathematical truth, for example, that a line is shorter than part of itself.

Statement: All lawyers are smart people.
Contrapositive: All people who are not smart are not lawyers.

Having proved the statement "If $x+5=7$, then $x = 2$" by proving "If $x \neq 2$, then $x+5 \neq 7$," you should not be surprised to be told that a statement and its contrapositive are logically equivalent. That is to say, they have the same truth value.

For any given conditional statement there is another closely related statement that you should know about, and it is called the **converse** of the given statement. The converse of a statement is created by switching the hypothesis and the conclusion. For example, the converse of the statement, "If Tabby is a cat, then Tabby has four legs." is "If Tabby has four legs, then Tabby is a cat." For the following statements, write the converse.

1. If you like pizza, then you must love Uno's.

2. If a triangle has two equal angles, then it has two equal sides.

3. All lawyers are smart people.

Considering the above statements, answer the question, "Is the converse of a true statement always true?"

The above examples have used almost exclusively the "If...then..." form of the conditional statement, but often you will need to recognize the converse of a statement that is in some other form. See if you can write both the contrapositive and the converse of the following statements.

4. If n is odd, then n + 1 is even.

5. All even numbers are divisible by 2.

See if you can write the converse of these next two statements.

6. If one angle of a triangle is greater than another, then the side opposite the greater angle is greater.

7. If Jim refuses to go to the football game, neither his brother nor his sister will be speaking to him.

Both the converse and the contrapositive are concepts that you will want to be familiar with, whether thinking about mathematics or other subjects. The important thing for you to know right now is that if a statement is true, then its converse may or may not be true, but that the contrapositive of a statement always has the same truth value as the original statement.

Teacher's Notes

At some point students may notice that Euclid, although he continually deals in lengths and areas, never measures anything. All his measurements are comparative. That is, he states that one line is longer than another, or that one figure is double another. While the reasons for this are beyond a high school mathematics course, at the least it can be pointed out that measurement is not allowed, for example, in the construction propositions. All constructions need to be done with a straightedge and compass, which by nature produce figures that conform (theoretically) to Definitions 4 and 15.

Notes on Book I

Definitions

- Have students try to figure out what the definitions mean rather than explaining them right away, especially the earlier definitions of point, line, surface, boundary, figure.
- Encourage students to notice the properties of things based on the definitions.
    - How many points do we know for sure are on a line?
    - Is the intersection of two lines a point?
    - Is a line made up of points?
    - Is a figure, like a circle, just the boundary, just what is inside the boundary, or both?
- Explain what "finite" and "infinite" mean.
    - What, on the earth, is infinite?
    - What is Euclid's purpose in introducing the term "infinite?"
- Before reading the definition of a right angle, have the students try to formulate their own definition; do the same for the definition of a circle
- Point out that triangles are classified two ways
    - by what kinds of angles they have (acute, obtuse, right)
    - by properties of their sides (equilateral, isosceles, scalene).
- Point out that the intersection of two sides of a rectilinear figure is called a **vertex**.

There is no need to go through the last two definitions until you are quite far along in Book I, but when you do be sure to discuss terminology that is different today than it was in Euclid's day. Define trapezoid; is this the same as trapezium? What is a rhomboid called today? What does Euclid call a parallelogram? etc.

Postulates and common notions

- Discuss the difference between postulates and common notions.
- Introduce the term "axiom."

168

- Mention that there are some common notions that are used that Euclid does not state.
    - the transitivity of "<" and ">"
    - equals added to or subtracted from an inequality preserves the inequality
    - the law of trichotomy for magnitudes, numbers, or lines: for any two comparable quantities x and y, one and only one of the following is true: $x < y$, $x = y$, or $x > y$
- Note that two figures being equal means equal in area (C.N.4), not shape.

Propositions

The first three propositions in Book I are constructions. In one way this is unfortunate, because the proof of a construction (relative to creating a drawing to work with) is different from that of a conditional statement to be proved and is sometimes puzzling to the student.

To illustrate the difference, consider first Proposition I.6: *"If in a triangle two angles equal one another, then the sides opposite the equal angles also equal one another."* One might start a proof by making a drawing of a triangle, labeling the vertices *A*, *B*, and *C*, and then saying something like, "Let *ABC* be a triangle with angle *ABC* equal to angle *ACB*. I say that *AB* = *AC*." This is quite straightforward. The hypothesis of the conditional statement, *"in a triangle two angles equal one another,"* forms the basis for making the drawing and the conclusion of the conditional statement, *"the sides opposite the equal angles also equal one another,"* forms the basis of the assertion we make about the drawing (*AB* = *AC*). It is then the job of the student to figure out, based on the drawing, how the proof goes.

By contrast, Proposition I.1 states, "To construct an equilateral triangle on a given finite straight line." Here the hypothesis is, "we are given a finite straight line," and the conclusion is, "we can construct an equilateral triangle on it." So we begin a proof with something like "Let *AB* be the given finite straight line. It is required to construct an equilateral triangle on the straight line *AB*." But this drawing gives us nothing (relative to itself) to prove. We have no triangle *ABC* about which we can say, "we need to show that *AB* = *AC* = *BC*." In order to prove we can construct an equilateral triangle we must figure out how to do it, make a drawing of what we do, and demonstrate as we go along that our steps are valid ones and lead us to the required construction. Logically, of course, there is no problem. But from a student's standpoint it is less satisfying, in that when one reads "I say …" it does not follow that there is a drawing to look at and try to figure out how the proof is going to proceed.

As mentioned above, all constructions are drawn using only a straightedge and compass, and students will need to have this explained.

Proposition I.1
- Point out the "if…then…" nature of propositions (students who have trouble with this concept can be directed to Appendix B)
- Explain that "Given:…To Prove…" is another form of stating a proposition.
- Use this proposition to explain what you expect in a proof ("translating" Euclid).
- Emphasize that the proof of a proposition begins with the creation of a drawing illustrating the enunciation, after which comes a statement of the proposition as it applies to the drawing. For example, Proposition I.1 states "To construct an equilateral triangle on a given finite straight line," so the first statements of Proposition I.1 would be:

"Let AB be the given finite straight line.
It is required to construct an equilateral triangle on the straight line AB."

The main body of the proof might look something like the following:

1) With center A and radius AB draw circle BCD — Post. 3
2) With center B and radius BA draw circle ACE — Post. 3
3) Draw lines AC and BC — Post. 1
3) $AB = AC$ — I. Def. 15 applied to circle ACD
4) $BA = BC$ — I. Def. 15 applied to circle ACE
5) $AC = BC$ — C. N. 1
6) Therefore, triangle ABC is equilateral — I. Def. 20
Q.E.F.

- Note that Euclid does not address the problem of how one knows that the two circles drawn in steps 1 and 2 will intersect, nor does he mention that there are two points of intersection of the two circles.

Proposition I.4
Upon studying this proposition, two questions come immediately to mind:

- No mention has been made in the definitions, postulates, or common notions, about "applying" one figure to another.
- If you can "apply" one figure to another, then why, in the proof of Proposition I.3, did Euclid not simply "apply" the given line to the given point and be done with it?

These questions cannot be answered satisfactorily, a fact undoubtedly known to Euclid, since, although there are a number of propositions where he might have devised a proof that "applies" one figure to another, he does not use that method unless there appears to be no other method of proof. Many people have thought that Proposition I.4 should have been a postulate. For those interested, Sir Thomas Heath has a detailed discussion of this proposition, in which he mentions the fact that geometers of Euclid's day apparently considered "applying" one figure to another a valid method of proof.

At this point in the school year students have not seen lots of proofs, so they may not notice either that this proof is different from others, or if they do notice they may not see any difficulty with it. If it seems appropriate, three difficulties worth discussing are that the proof

- "applies" one triangle to another
- does not appeal to definitions, postulates, or common notions
- assumes the converse of Common Notion 4

Congruence (a term not used by Euclid but one they will encounter in any modern text) can be introduced at this point.

Proposition I.6
- Introduce and explain the concept of an indirect proof before doing this proposition
- Define "converse" and point out that Proposition I.6 is actually the converse of part of Proposition I.5.

Proposition I.8
- Note that this is another proposition that establishes congruence.
- Introduce abbreviations SAS and SSS if it seems appropriate.

Proposition I.9 – I.12
- After getting through these propositions, there is a practice assignment in doing constructions. It consists of the following construction and others similar to it.

  Given a triangle, carefully construct the perpendicular bisectors of the sides and see where they meet.

Of course, the idea is to observe that each set of three lines appears to intersect in a single point.

The following two suggested exercises are similar to the construction exercises in the book. The student is asked to construct a figure, and upon doing so, the student observes that the figure he constructed appears to have a certain property. But when he attempts to prove the figure actually has this property, he seems unable to do so.

- Have the students construct a square by erecting perpendiculars and then have them try to prove it is a square. (This cannot actually be proved until after the propositions on parallel lines.) Have them try to figure out exactly what goes wrong when they try to prove they have constructed a square.

- Have the students bisect a straight line, and on each section of the line, on the same side of the line, construct an equilateral triangle. Let them connect the remaining points of each triangle and then let them try to prove that the middle (inverted, as it were) triangle is also equilateral. (This cannot actually be proved until after Proposition I.32.)

Proposition I.14 Point out how Proposition I.14 is a converse of Proposition I.13

Proposition I.16 Explain what an exterior angle is relative to a convex figure. Notice what happens as the number of sides of the figure increases. See note on Proposition 32.

Propositions I.19 and I.20 Another set of converses

Proposition I.26 This is a third (and fourth really) proposition that establishes congruence. Discuss its abbreviations (AAS, ASA) and ask if they think AAA or SSA might be true.

Proposition I.27
- Point out to the student how to figure out from the text (if they don't know simply by the definition of the words) what is meant by "alternate," "interior," "exterior," etc. angles.
- Review the definition of parallel lines.

Proposition I.32
- After they have proved this proposition give the students the drawing below and ask them if they can discover a proof of the second part of this proposition based on this drawing.

- Discuss with the students again (thinking back to Proposition I.16) exterior angles. Now that they know the sum of the angles of a triangle is two right angles, ask them to figure out the sum of the interior angles of rectilinear figures with varying numbers of sides. Once they know what the sum of the interior angles of an n-sided rectilinear figure is, have them figure out the sum of all the exterior angles of an n-sided figure.

Proposition HL (In the set of exercises after Proposition I.32) Note that this proposition is a special case of SSA and remind the students that SSA is not in general true.

Proposition I.34 Note that Euclid uses a different definition of parallelogram than the one he gives in Definition I.22.

Proposition I.43 Note new definitions – complement, parallelogram "about" a diameter.

Proposition I.47 and I.48
- Less skilled students have an easier time following the proof if it is broken up into parts.
- Note that we have another set of converses in Propositions I.47 and I.48, but that Proposition I.47 is the one called the Pythagorean Theorem.

Notes on Book II

Book II, for the most part, is quite different in tone from Book I. The proofs of the first ten propositions do not build upon one another at all but are independent, each depending predominantly on the propositions in the latter part of Book I. Because the technique of proof is the same for all of them the student who does not understand the purpose of the propositions may feel that they are repetitious yet still difficult, while the student who grasps the underlying purpose of them may find the repetition a bit trying because it seems obvious. The former student, of course, must be encouraged to persevere in his efforts to see the underlying principles at work in these propositions. And fortunately there is a challenge you can give the latter student to force him to think hard about these propositions.

In point of fact, the first ten propositions are geometric demonstrations of common algebraic identities. To give two examples, Proposition II.1 is the geometric equivalent of $x(y_1 + y_2 + ... + y_n) = xy_1 + xy_2 + ... + xy_n$, while Proposition II.4 is the geometric equivalent of $(x+y)^2 = x^2 + 2xy + y^2$. Even your best students are likely, especially at first, to find it a challenge to decide what algebraic truth is represented by each of the geometric propositions. To do so is a good exercise, since it creates an intuitive connection between algebra and geometry, which later on, if the student continues in his study of mathematics, may be made formal.

One interesting aspect of looking at algebraic equivalences from a geometric standpoint is that at times it is possible to demonstrate the geometric equivalence of one algebraic statement only by considering two cases, even though there is no need in algebra for two separate statements. Propositions 5 and 6, Propositions 9 and 10, as well as Propositions 12 and 13, are examples.

Propositions II.5 and II.6 are the first propositions where you can think of what happens if one thinks in terms of the direction of lines being positive or negative (which would lead to the possibility of area being negative). If you visualize the point D in the drawing for Proposition II.5 being "slid" to the right until it passes point B (so that in one case the "length" BD is positive and in the other it is negative), then your drawing will become the drawing of Proposition II.6 and it is not hard to see that the two propositions are really describing the same thing algebraically, even though geometrically the two cases must be considered separately (because, in fact, we don't have negative lengths).

Propositions 9 and 10 See previous comment (for Propositions II.5 and II.6).

Proposition II.11 is the proposition that is used to prove Proposition IV.10, which then allows one to complete Book IV. Without covering Proposition II.11 one cannot move past Proposition 9 in Book IV.

Propositions II.12 and II.13 are both geometric equivalences of the law of cosines (and as such are generalizations of the Pythagorean Theorem). However, as in Props II.5 and II.6 and Props II.9 and II.10, the geometric proof must consider two cases to be complete.

The following table gives the algebraic equivalences of the first ten propositions of Book II.

Prop	Statement	Algebraic Equivalent
II.1	If there are two straight lines, and one of them is cut into any number of segments whatever, then the rectangle contained by the two straight lines equals the sum of the rectangles contained by the uncut straight line and each of the segments.	$x(y_1 + y_2 + \ldots + y_n) = xy_1 + xy_2 + \ldots + xy_n$
II.2	If a straight line is cut at random, then the sum of the rectangles contained by the whole and each of the segments equals the square on the whole.	$x^2 = xy + x(x - y)$
II.3	If a straight line is cut at random, then the rectangle contained by the whole and one of the segments equals the sum of the rectangle contained by the segments and the square on the aforesaid segment.	$xy = y(x - y) + y^2$
II.4	If a straight line is cut at random, then the square on the whole equals the sum of the squares on the segments plus twice the rectangle contained by the segments.	$(x + y)^2 = x^2 + 2xy + y^2$
II.5	If a straight line is cut into equal and unequal segments, then the rectangle contained by the unequal segments of the whole together with the square on the straight line between the points of section equals the square on the half.	$(x + y)(x - y) + y^2 = x^2$
II.6	If a straight line is bisected and a straight line is added to it in a straight line, then the rectangle contained by the whole with the added straight line and the added straight line together with the square on the half equals the square on the straight line made up of the half and the added straight line.	Same as II.5
II.7	If a straight line is cut at random, then the sum of the square on the whole and that on one of the segments equals twice the rectangle contained by the whole and the said segment plus the square on the remaining segment.	$x^2 + y^2 = 2xy + (x - y)^2$
II.8	If a straight line is cut at random, then four times the rectangle contained by the whole and one of the segments plus the square on the remaining segment equals the square described on the whole and the aforesaid segment as on one straight line.	$4xy + (x - y)^2 = (x + y)^2$
II.9	If a straight line is cut into equal and unequal segments, then the sum of the squares on the unequal segments of the whole is double the sum of the square on the half and the square on the straight line between the points of section.	$(x + y)^2 + (x - y)^2 = 2(x^2 + y^2)$
II.10	If a straight line is bisected, and a straight line is added to it in a straight line, then the square on the whole with the added straight line and the square on the added straight line both together are double the sum of the square on the half and the square described on the straight line made up of the half and the added straight line as on one straight line.	Same as II.9

Notes on Book III

Definitions

Definition 2 Explain that "is tangent to" is the modern equivalent of "touches."

Definition 6 Explain that "arc" is the modern word for what Euclid calls "circumference." If this usage of "circumference" to designate a portion, rather than the whole, seems odd, recall that we do the same for portions of lines (calling them "lines" as well) with no confusion.

Definition 7 I suggest omitting the definition of an angle of a segment, since for a truly adequate definition you will get into the concept of a limit. You can explain why it is omitted if students are interested and can absorb it. The only difficulty is that the second half of Proposition III.16 will have to be omitted.

Definition 11 This definition really doesn't make sense until after Proposition 21 when it is proved that angles in the same segment are equal.

Propositions

General comment: As always, all propositions are crucial but some are used more than others. In Book III the ones most often referred to (in other propositions and in problem sets) are Proposition 1 and its corollary and Propositions 3, 9, 14, 16, 18, 19, 20, 21, 22, 31, and 32.

Proposition III.1
Note that line AB is arbitrary and that Proposition 1 proves that the center of the circle lies on the perpendicular bisector of line AB. Develop a second way of finding the center of a circle. With help students will usually come up with drawing another line *CD* in the circle and drawing the perpendicular bisector of that line. They will see that the point of intersection of the two lines must be the center of the circle. Ask them if it is possible for this construction to get them into trouble, i. e., is some way to draw the two lines so that the perpendicular bisectors do not identify the center. After they figure out that if *AB* and *CD* are parallel to one another then their perpendicular bisectors are the same line, ask them how to solve the problem. Eventually someone will realize that if the two lines have a common endpoint they cannot be parallel.

Discuss how many points determine a circle. You may be surprised at how long it takes them to decide that an infinite number of circles can be drawn through two given points. Go to four points and ask how many circles can be drawn through four given points. After they decide that three is probably the number of points it takes to determine a circle give them three points and ask them to figure out how to find the center and radius of the circle that passes through those three points. In spite of having discovered the alternate way of finding the center of a given circle they will most likely still have difficulty realizing what to do with the three given points.

This is a good time to give a quiz to solidify definitions, terminology, and constructions.

Proposition III.2
For extra credit ask if anyone can come up with a direct proof of Proposition III.2.

Propositions III.5 and III.6
Point B in the drawing accompanying Proposition 5 is never used. If you change the letters in Proposition III.6 to match those of Proposition III.5, the two proofs are identical.

Propositions III.7 and III.8
These propositions are not difficult to understand but they are lengthy. They are not referred to again in any future propositions. However, the proof of Exercise 1 on page 55 depends on Proposition III.7.

Proposition III.9
This proposition, of course, can be related to the discussion after Proposition III.1 that focused on finding the center of a circle.

Proposition III.14 At this point it is a good idea to review Definitions 4 and 5.

Proposition III.16
The second half of this proposition deals with an "angle of a semicircle." The second part of this proposition can be omitted if the notion of an angle between a straight line and an arc has not been defined. Of course, one needs to observe that the line constructed at the endpoint of a diameter is tangent to the circle.

Proposition III.16, III.18, and III.19
These three propositions relate the concepts of a line touching a circle, a line being perpendicular to a radius at its extremity, and a line passing through the center of a circle. If any two of these conditions are satisfied, then the third is also.

Proposition III.21 After this proposition is proved, Definition 11 should be studied.

Proposition III.22
There seems to be a strong temptation for students to forget the hypothesis of this proposition and to come to believe that in any quadrilateral the opposite angles equal two right angles.

Propositions III.23 – III.26 These are related and quite straightforward.

Notes on Book IV

Book IV of the *Elements* deals exclusively with constructions. Euclid first defines "inscribed" and "circumscribed" relative to circles and polygons and then proceeds, first with triangles and then with regular polygons of 4, 5, 6, and 15 sides, to do the following:

- inscribe a polygon within a circle,
- circumscribe a polygon about a circle,
- inscribe a circle within a polygon, and
- circumscribe a circle about a polygon.

For triangles, the Exercises on Constructions in Book I laid the groundwork for doing the last two of these constructions (Propositions IV.4 and IV.5), so these are not unfamiliar to the student, even though the proofs may be.

For squares (Propositions IV.6 – IV.9) the constructions are particularly easy, requiring only some of the most intuitive and frequently used propositions from Book I (mostly relating to parallel lines) and the equally intuitive and frequently used propositions from Book III about the perpendicularity of a line tangent to a circle at the point of tangency. For this reason these constructions have been placed in a special exercise set in between Books III and Book IV and it is hoped that students will do the exercises without realizing that they are actually propositions from the next book. It goes without saying that there are many ways of proving these propositions, and students' proofs can be expected to vary considerably.

For pentagons, unfortunately, the situation is not so simple. The method Euclid used to inscribe a regular pentagon within a circle requires that he first find a way to construct an isosceles triangle, the base angles of which are each twice the remaining angle. He does this in Proposition IV.10, and it is this proposition that requires the use of Book II, specifically Proposition II.11. If Book II is skipped, then one must basically end the study of Book IV at Proposition IV.9. Proposition IV.15 (hexagons) can be covered, but that is all.

(Of course, if you and your class can accept on faith that Proposition IV.10 is true, then it is possible to finish Book IV.)

Notes on Book V

Definitions

In Book V Euclid introduces the concepts of ratio and proportion, restricting his focus to ratios and proportions of lines. He develops the definitions and rules for working with ratios and proportions as preparation for Book VI, which contains all the familiar propositions about proportionality within and between rectilinear figures as well as some that are not so familiar.

Definition 5, which defines equality of ratios, may be a puzzle at first glance, until one remembers that at this point in the *Elements* number theory has not been explored. In fact, the concept of number has yet to be introduced. A ratio has been defined simply as a relation in respect to size between two magnitudes of the same kind. So Euclid, when he wishes to compares lines, as in a ratio, does not compare lengths of lines by equating them with numbers, something we might encounter in a modern text. Rather, he compares them by making reference to some measure that the lines share, saying, for example, that line A is twice line C and line B is three times line C. The ratio of line A to line B is then 2 to 3.

The problem, of course, is that not all lines share a common measure. The side of a square and its diagonal are one such example. Such lines are called incommensurable. Now obviously Euclid's definition of ratio could not be restricted to lines that were commensurable. He needed to be able to compare any two lines, even incommensurable ones, in any two rectilinear figures. So he needed a definition of equality of ratios that did not depend upon whether or not the lines in question were commensurable. The definition he came up with is quite ingenious.

A careful reading reveals the following. There are four magnitudes under consideration, which are called the first, second, third, and fourth magnitudes. For the purpose of explaining what is happening we are going to consider the magnitudes as if they were numbers (rather than lines) and we are going to express ratios in fractional form. (Mathematically, of course, we are on safe ground, but we do not have any justification at this point in the *Elements* for doing this.)

Now if we call the four magnitudes a, b, c, and d, and express ratios in fractional form, then the first and third magnitudes of which Euclid takes equimultiples are the antecedents of two ratios, and the second and fourth magnitudes are the consequents. Thus, we want to decide what it means, according to Euclid's definition, to say that a is to b as c is to d (or $a:b = c:d$). Equivalently, we want to decide what it means to say

$$\frac{a}{b} = \frac{c}{d}$$

Euclid says two ratios are equal if, any time you take any multiple whatever of the first and third magnitudes (the numerators) and any, possibly different, multiple of the second and fourth magnitudes (the denominators), then whatever the resulting relationship between the numerator and denominator of the first fraction, whether the numerator is less than, or equal to, or greater than, the denominator, that same relationship will hold for the second fraction. That is, looking at the fractions

$$\frac{ma}{nb} \text{ and } \frac{mc}{nd}$$

where m and n are any magnitudes whatever, if ma is less than nb then mc will be less than nd, and if ma equals nb then mc will be equal to nd, and if ma is greater than nb then mc will be greater than nd. If that is the case for any choice of multiples, then the ratios a:b and c:d are equal.

A very simple example using integers will demonstrate what happens when the ratios of two pairs of integers are not equal. Suppose the ratio a:b is 3:2 and the ratio c:d = 4:3. To see that these two ratios are not equal, that is, to see that

$$\frac{3}{2} \neq \frac{4}{3}$$

we will show that Euclid's definition is not satisfied.

If we choose the multiple for the numerators (the first and third magnitudes) to be 2, and the multiple for the denominators (the second and fourth magnitudes) to be 3, then the resulting ratios to be compared are

$$\frac{2(3)}{3(2)} \text{ and } \frac{(2)4}{(3)3}$$

Since the numerator and denominator of the first ratio are equal, but in the second ratio, the numerator is less than the denominator, the two original ratios are unequal.

In a case involving incommensurate lines, of course, it is impossible to achieve equality by taking equimultiples, but in that case, with a proper choice of multiples, the antecedent of one ratio can be made less than the corresponding consequent while in the other ratio the situation will be reversed. A little experimentation with numbers will show that if you are comparing any two unequal ratios and wish to show Euclid's definition is not satisfied, all you need to do is find a fraction, say n/m, that is between the values of the two ratios (expressed as fractions) that you are comparing. Using m as the multiple for the two numerators and n as the multiple for the two denominators, when you perform the required multiplications you will have in one case the numerator larger than the denominator and in the other case the numerator smaller. No effort to prove that will be made here but it is not hard to demonstrate.

For those interested, in Sir Thomas Heath's translation of the *Elements* there is a somewhat lengthy, but very good illustration, devised by DeMorgan and using lines, of how this definition works. For those who have the background, there is also an excellent discussion of the correspondence "almost coincidence," as Heath says, between Euclid's definition of equal ratios and Dedekind's theory of irrational numbers.

Definitions 9 and 10 are also definitions, which may give difficulty as expressed, although the algebraic results they lead to are quite familiar. Definition 9 says that if three magnitudes are proportional, then the first is said to have to the third the **duplicate ratio** of that which it has to the second. In other words, if $a{:}b = b{:}c$, then $a{:}c$ is called the duplicate ratio of $a{:}b$. Put algebraically, $a{:}c$ is the duplicate ratio of $a{:}b$ provided

$$\frac{a}{b} = \frac{b}{c}$$

For an obvious example, 9:1 is the duplicate ratio of 9:3 (equivalently, 3:1) since

$$\frac{9}{3} = \frac{3}{1}$$

But suppose you have been told that some ratio $x{:}y$ is the duplicate ratio of another ratio $a{:}b$. What does that mean? Suppose, for example that $x{:}y$ is the duplicate ratio of 3:2. To know what x and y are you need to find the duplicate ratio of 3:2. That is, you need to find a number c, satisfying

$$\frac{3}{2} = \frac{2}{c}$$

Solving for c, we find that $c = 4/3$. So the three magnitudes, 3, 2, and 4/3 are proportional and therefore 3:4/3 is the duplicate ratio of 3:2. But 3:4/3 is more simply written as 9:4. Thus, 9:4 is the duplicate ratio of 3:2.

We can obviously follow the same procedure for the general case. In order to find $x{:}y$, the duplicate ratio of $a{:}b$, we must find another quantity c, satisfying the equation

$$\frac{a}{b} = \frac{b}{c}$$

But solving this equation for c, we find that

$$c = \frac{b^2}{a}$$

Therefore, to say $x{:}y$ is the duplicate ratio of $a{:}b$ is to say

$$\frac{x}{y} = \frac{a}{\frac{b^2}{a}} = \frac{a^2}{b^2}$$

Hence, the duplicate ratio of $a{:}b$ is the ratio $a^2{:}b^2$. Similarly, it can be shown that the triplicate ratio of $a{:}b$ is $a^3{:}b^3$.

Definition 12 - Note that this definition can apply only when all four magnitudes are of the same kind. So when, in Book VI, we learn that triangles under the same height are to one another as their bases, we cannot use Definition 12 and say that for two triangles under the same height, the first triangle is to its base as the second triangle is to its base. A triangle and its base are not the same kind of magnitude, one being an area and the other being a line.

Propositions

Propositions V.1-V.3 and V.5 and V.6 These are not really about ratios, but demonstrate common properties relating multiples and magnitudes. For example, Proposition V.1 states that multiplication by a multitude distributes over addition of magnitudes. For convenience we list them here. As usual, m and n are multiples; x's and y's are magnitudes.

Proposition	Statement
V.1	$m(x_1 + x_2 + ... + x_n) = mx_1 + mx_2 + ... + mx_n$
V.2	$(m+n)x = mx + nx$
V.3	$m(nx) = (mn)x$
V.5	$m(x-y) = mx - my$
V.6	$(m-n)x = mx - nx$

Many of the propositions of Book V are simply common properties of ratios, proved as they relate to lines. The first column of the following table is a list of the most commonly used propositions regarding ratios and proportions. The second column expresses the given proposition symbolically. And the third column (if filled in) expresses the given proposition in modern language. As usual, m and n represent multiples, while the remaining variables represent magnitudes.

V.4	If $w:x = y:z$, then $mw:nx = my:nz$	
V.7	If $x = y$ then $x:z = y:z$ and $z:x = z:y$	Equal magnitudes are in equal ratios to a third.
	Corollary* If $w:x = y:z$, then $x:w = z:y$	Inverse proportion
V.9	If $x:z = y:z$, then $x = y$	Two magnitudes which have the same ratio to a third magnitude are equal.

	and if $z{:}x = z{:}y$, then $x = y$	If a magnitude has the same ratio to two magnitudes then the two magnitudes are equal.
V.11	$x{:}y = z{:}w$ and $z{:}w = u{:}v$ then $x{:}y = u{:}v$	Transitivity of ratios
V.12	If $x_1 : y_1 = x_2 : y_2 = ... = x_n : y_n$ then $(x_1+...+x_n):(y_1+...+y_n) = x_i : y_i$ for i between 1 and n	For EQUAL ratios, the ratio of the sum of the antecedents to the sum of the consequents is equal to any one of the original ratios.
V.14**	If $w{:}x = y{:}z$ and $w > y$, then $x > z$	
V.15	$x{:}y = nx{:}ny$	Parts have the same ratio as the same multiples of them.
V.16**	If $w{:}x = y{:}z$, then $w{:}y = x{:}z$	Alternate proportions
V.22	If $x_1 : x_2 = y_1 : y_2, x_2 : x_3 = y_2 : y_3, ..., x_{n-1} : x_n = y_{n-1} : y_n$ then $x_1 : x_n = y_1 : y_n$	Ratio ex aequali
V.25**	If $x{:}y = z{:}w$, then the greatest and the least together are more than the other two	The sum of the extremes is greater than the sum of the means.

*This corollary, while certainly true, does not follow from V.7, but follows directly from the definition of two ratios being equal.

** This proposition only makes sense (in light of the Book V definitions) if all four magnitudes are of the same kind.

Notes on Book VI

Book VI contains the familiar propositions regarding relationships of proportionality or equality between elements of similar figures. Even though the text stops at Proposition 13, the exercises take the students through Proposition 16. And the last exercise has the students prove Proposition 19. The propositions in the exercises are stated in a way to maximize the chances that students will be able to prove them on their own, and in the later exercises parts of the propositions are in outline form, asking the student to "flesh out" the proof. It is, of course, not recommended that students be told that these exercises are future propositions.

Before beginning Book VI the student needs to be familiar with the word "contained" as applied to rectangles. That is, a rectangle is said to be **contained** by the two straight lines which form a right angle.

Definitions

Definition 1 – This definition is peculiarly (and inadequately) worded. Of course, for figures to be considered similar the angles in the two figures must be equal in the same order and the corresponding sides must then be proportional.

Definition 2 – Reciprocally related figures is defined here in terms of reciprocally proportional sides, which, unfortunately, has not been defined. But it is clear from its usage what it must mean.

Propositions

Propositions VI.6 and VI.7 – these correspond to congruence Proposition I.4 (S.A.S) and what would be a S.S.A. if there were such a proposition.

Notes on Selected Exercises

Page 17, Exercise 7

Most students know that the angles of a triangle sum to two right angles. (They learned this as 180) but this fact has not yet been proved in the *Elements*. The objective of the exercise is to see if they can realize, as they struggle to prove the obvious counterexample, what it is that they are missing.

Beware of a "proof" that relies on the visual, or that misuses Proposition I.3.

Page 37, Exercise 4

Draw line AC. Then apply Proposition I.32 to both $\angle FBC$ and $\angle GDC$, do a little addition, and the job is finished.

Page 38, Exercise 10

This requires an indirect proof. Assuming CE is not less than ED leaves two cases to consider; either $CE = ED$, or $CE > ED$.

$CE = ED$ results in $\triangle AEG \cong \triangle BED$, a contradiction of the fact that $\angle CAE < \angle DBE$.

$CE > ED$ allows the placement of point G on EC in such a way that $GE = ED$. This leads to $\triangle GAE \cong \triangle DBE$, resulting in $\angle GAE$ being simultaneously less than $\angle CAE$ and equal to $\angle DBE$, a contradiction since $\angle DBE$ was given as greater than $\angle CAE$.

Page 55, Exercise 1

Note that $AP = PC$ and that $PE = PE$. That is, $\triangle APE$ and $\triangle CEP$ have two sides equal. In order to prove, then, that $\angle APE > \angle CPE$, one need only prove $EC < AE$. Now note that $PE \perp BD$ (Proposition III.3). Therefore, $EC \neq AE$ (Proposition III.4). Demonstrate that EC is shorter than AE (use Proposition III.7) and the proof is complete.

Page 56, Exercise 5

The simplest proof is to draw a circle concentric with circle $ABCD$ and tangent to BD at P. Then draw another line through P. Regarding the smaller circle, you can conclude that the distance from this new line to the center is less than the radius of this small circle. But looking at the larger circle, this means that the new line is longer than BD (Proposition III.15).

Page 66, Exercise 2

Use Proposition III.31 to construct a right angle within a circle, and with the vertex (of the right angle) as center, construct a circle of radius less than the length of the shortest leg of the right triangle. Apply Proposition III.20 to this new circle to create an angle that is half a right angle. You can continue this process (theoretically) indefinitely.

Page 111, Exercise 1 - This exercise can easily be done using only Propositions VI.1 and VI.2.

Page 111, Exercises 4-7 – These four exercises are statements of the results of Propositions VI.14 and VI.15 (each consisting of a statement and its converse). The drawing has been made for the student for Exercises 4 and 5, but in Exercises 6 and 7 the student is asked to extend his learning and create the proposition, the drawing, and the proof. The process follows that of Exercises 4 and 5 and should not be difficult.

Page 111, Exercise 4 – The hint in the exercise is to add a couple of lines. Obviously, it is lines *AF* and *CE* that need to be extended so that Proposition VI.1 can be applied to the two overlapping sets of parallelograms.

Page 111, Exercise 8 – This exercise is the geometric equivalent of what is commonly called "cross multiplication" in algebra. Presuming that the class will go on to introduce measurement and use the results that have been proved throughout the year, we have asked the question, "This proposition has a very common algebraic counterpart. What is it?" It will probably be a revelation when they realize what they have done.

Page 111, Exercise 10 – This is the familiar result that similar triangles are to one another as the square of corresponding sides. The proof, however, does not make that clear, and later on, when the class is involved in problem solving, the algebraic meaning of this proposition can be explored (See notes on Book V, Def 9).

Made in the USA
Middletown, DE
08 July 2020